The Physical Environment and Behavior

AN ANNOTATED BIBLIOGRAPHY AND GUIDE TO THE LITERATURE

The Physical Environment and Behavior

AN ANNOTATED BIBLIOGRAPHY AND GUIDE TO THE LITERATURE

Joachim F. Wohlwill
and
Gerald D. Weisman

The Pennsylvania State University
University Park, Pennsylvania

PLENUM PRESS · NEW YORK AND LONDON

Library of Congress Cataloging in Publication Data

Wohlwill, Joachim F.
   The physical environment and behavior.

   Includes index.
   1. Environmental psychology — Bibliography. I. Weisman, Gerald D.   II. Title.
Z7204.E55W63 [BF353]   016.3042   81-4840
ISBN 0-306-40739-6   AACR2

© 1981 Plenum Press, New York
A Division of Plenum Publishing Corporation
233 Spring Street, New York, N.Y. 10013

Printed in the United States of America

ACKNOWLEDGMENTS

The preparation of an extensive bibliography such as this
volume inevitably involves the collaborative efforts of a
considerable number of persons working in a variety of
capacities. Among these, we wish to acknowledge in
particular the assistance of the following individuals:

Don Hazle, who set up the program for the computer-based
typescript and gave freely of his time and advice in the
preparation of the ms. as it went through its successive
stages.

Norma Eckenroth and Arlene Lear, who worked diligently
and expertly on the typing of the ms., shepherded it through
its several revisions, and produced the final text.

Jean Shorett, who participated in the early phases of the
planning of this bibliography, and contributed annotations
for Chapter 5.

Diane Jimski and Nitza Vega who provided bibliographic
assistance in the compilation of the material for Chapters
10 through 13.

Suzy Weisman, who did the painstaking work of copy-
editing the ms.

Terry Ramanujam, who assisted in the final editing of the
ms. and prepared the index.

To all of these persons -- and any others whom we may
inadvertently have omitted -- our sincere thanks. Last, but
not least, we owe a particular debt of gratitude to the
Program in Man-Environment Relations at the Pennsylvania
State University for its support, moral and material, of
this project.

September, 1980

J. F. W.
G. D. W.

CONTENTS

CONTENTS

INTRODUCTION

The field of "Environment-and-Behavior"

This bibliography is aimed at the researcher and advanced
student working in the field of environmental psychology, as
it has come to be designated over the past decade. A more
appropriate term might be "environment-behavior studies," to
suggest the important characteristic of this field as one
that transcends the province of the psychologist, and brings
together workers, as well as problems, methods, and concepts
from a great diversity of disciplines and professional
fields. Among these we may include geography and sociology,
architecture, landscape architecture and planning, forestry,
natural resource management and leisure and recreation
research -- to name only the most important of the diverse
fields from which material for this bibliography has been
drawn.

This is in fact one of the primary reasons for our belief
in the value of such a volume. The literature in the
environment-behavior field is scattered through the most
diverse sources, including not only the major periodical and
monographic literature in each of the above-mentioned
disciplines and professions (and others as well), but also a
variety of more specialized publications of varying degrees
of accessibility. Thus it seemed to us helpful to the
researcher, teacher and student in this area to bring this
far-flung literature together in a single volume, that might
be used as a guide to the field. We aimed at a
comprehensive treatment, including both basic and applied
aspects, and relations of behavior both to the man-made or
artificial and to the natural environment.

A brief overview of the field as we see it may be
helpful, to indicate our perspective on it. Environmental
psychology as a subspecialty of psychology proper came into
being in the late sixties. This occurred in part at the
instigation of architects and planners, seeking the aid of
the behavioral scientist for guidance in the design of

buildings, (particularly those housing social institutions),
so as to enhance the satisfaction and well-being of their
users. Psychologists generally came into the field with
less of an applied, problem-solving orientation, and
approached it rather in terms of an interest in certain
theoretical and empirical issues. Thus, social
psychologists became "environmental" in terms of a focus on
spatial aspects of interpersonal behavior, while perceptual
and cognitive psychologists became interested in people°s
perception of different features of the large-scale
environment, and in their mental representation of it.
Interest in this latter problem came likewise from the side
of geographers, as well as from a number of theoretically
inclined architects and planners. Still other psychologists
saw in the field an opportunity to study the affective and
motivational response of the individual to stimulation, as
well as processes of adaptation and coping to conditions of
environmental stress, or to develop conceptions of attitude
and personality based on a concept of "environmental
dispositions." Finally, advocates of certain idiosyncratic
theoretical views and approaches to the study of behavior
found in environmental psychology a new realm for the
application and extension of their ideas. Foremost among
these are the behavioral ecologists, whose ecological
perspective and emphasis on the role of behavior settings as
primary determinants of behavior made them inherently
sympathetic to an environmental focus. In a rather
different vein we find the proponents of behavior
modification becoming involved in the domain of
environmental problems deriving from or aggravated by
individual behavior, thus providing them with a new testing-
ground for their concepts and methodologies.

This is necessarily a brief and most incomplete picture
of the diverse directions from which psychologists have
entered this field. The diversity is hardly surprising,
once we recognize that environmental factors cut across all
of our common categories of behavior, and relate to
behavioral problems at all levels, from the perceptual to
the motivational, from the cognitive to the interpersonal.
This very diversity may, ironically, have impeded the wider
acceptance of environmental psychology as a valid specialty
within the discipline of psychology -- in a very real sense
all psychology is "environmental," that is, interested in
the role of environmental forces in behavior. Nevertheless,

environmental psychologists, and courses and texts devoted
to the field, have clearly established themselves on the
psychological scene, and so a bibliographic reference work
devoted to their diversified concerns should prove of value.

But, as already indicated, the environment-and-behavior
field is the province of workers representing many
disciplinary and professional concerns, in addition to
psychology. Thus researchers from such applied specialties
as landscape architecture, as well as forestry and natural-
recreation have become involved in studies of the dimensions
of evaluative response to the environment and of preference
or choice of particular locales for recreation and other
forms of behavior; those in the field of transportation have
been interested in determinants of transportation-mode
choice, and of response to alternative transportation forms.
Similarly, architects and designers have studied such
questions as determinants of residential satisfaction, and
evaluation of buildings and facilities by their users. We
have tried, through the coverage provided by this volume, to
reflect this multifaced character of the field, by including
sections dealing with various applied concerns that have
sprung up over the past decade and which occupy an integral
place in the domain of environment-behavior studies,
alongside those of the environmental psychologists
identified primarily with the discipline of psychology.

Organization, content and format of the volume

The overall organization of the volume reflects the dual
character of the field alluded to above; of its four major
sections, the two principal ones are devoted to work on
basic psychological processes on the one hand, and to work
on applied problems on the other. Within one or the other
of these will be found the major categories of research that
have evolved in the environment-and-behavior field. It is
apparent that any division into basic vs. applied is far
from clear-cut, and there are numerous instances in which we
found ourselves in doubt as to which section a particular
study should be included in. In general we have handled
such issues of overlap through cross-referencing from one
section to another, by listing, at the end of each, a set of
item numbers referring to works that we considered relevant
to the subject-matter of that section, but which are
assigned to other chapters of the volume (or to other
sections of the same chapter).

We believe we have included in our coverage most if not
all of the significant topics to which those in the
environment-and-behavior field have addressed themselves.
One topic that some may miss in Part III on applied
problems, is that of Work Environments; likewise the
experimental work in the human factors area, much of which
relates to the role of specific environmental variables in
human task performance, has received only limited coverage
in this volume. Perhaps partly because of the rather narrow
focus of much of this work, this literature has remained
somewhat apart from the mainstream of environment-behavior
theory and research. This fact, as well as our lack of
expertise in these fairly specialized areas, motivated us to
exclude these topics from our volume. Within each of the
chapters, our coverage necessarily varies in
comprehensiveness and completeness, depending on such
factors as the sheer volume of material that has appeared on
any given topic, the accessibility of the literature in it
and its overall quality, and, realistically, our own
acquaintance with the area, and ability to retrieve the
requisite information. We have generally restricted
ourselves to material in books and journals that form part
of the published literature, and thus receive reasonably
wide distribution, or at least are accessible through major
research libraries, inter-library loan services, etc. This
meant that, with occasional exceptions, we have not included
material in the form of reports of governmental agencies,
research centers, and institutes, and similar types of
publications receiving only limited circulation, and
generally not regarded as part of the scholarly literature
on a given subject. Here, too, the dividing line is a thin
one on occasion, and exceptions have been made to allow us
to include a limited number of governmental and other in-
house publication whose importance to workers in a given
field appeared to warrant it. We have also included
selected dissertations, listed without annotations at the
end of each section, and referenced to their abstract in
Dissertation Abstracts International, which is readily
available in research libraries, and which provides
requisite information for ordering of theses themselves
through University Microfilms.

Further, we have for the most part confined ourselves to
material falling into one of three types: (a) empirical
studies dealing with some clearly specified aspect of

individual or group behavior, as related to some
identifiable set of environmental variables, or
environmentally-relevant issues; (b) material of a
theoretical, methodological or literature-review nature
relating to environment-behavior issues; and (c) essays,
discussions, and reviews with an explicit behavioral focus
on aspects of the natural or the man-made environment. This
meant that we excluded such material as discussions of urban
problems without a clear behavioral component, speculations
on the aesthetics of architecture or landscape, or on the
impact of the natural environment, formulated in terms that
are not readily analyzable in behavioral-science terms, and
other similar material that seemed to be of marginal
interest from the perspective of environment-behavior
research. A further limitation of our coverage, without
which the subject-matter dealt with in the volume would
appear even more amorphous and ill-defined than is presently
the case, is our restriction on material that deals with the
physical, as opposed to the social or institutional
environment. In this context, we disagree strongly with the
often-voiced view that it is not possible or profitable to
differentiate between the physical and the social
environment. It would be a quixotic world indeed in which
people respond to wind-mills as to other human beings, and
for the behavioral scientist to fail to observe this
fundamental differentiation between the world of people and
the world of objects, or of dimensions of the geographic
environment, is to court theoretical confusion and practical
disaster.

     As regards format, the following points should be noted.

     a. Annotations have of necessity been kept brief, and
non-evaluative. In the case of reports of empirical
studies, we have generally included reference to the chief
finding or findings.

     b. A number of readers and collections of edited papers
are listed in Chapter 2 Texts and Readers. For the most
important ones of these, made up of original (rather than
reprinted) papers, a complete table of contents is given;
for the others, we have limited ourselves to brief
descriptive annotations indicating the type of material
included. In both cases individual selections have been
selectively chosen for inclusion under the appropriate

topical headings; these have been referenced to the volume
in which they appear by the number of the item corresponding
to that volume in our bibiography. To facilitate
identification we have added the name of the editor(s), or,
in the case of items from the Proceedings of the
Environmental Design Research Association Conferences, the
"EDRA" acronym followed by the number designating the
Conferences. (Thus the paper by G. McKechnie, "Simulation
techniques in environmental psychology," is listed in the
section on Methodology as item #99 with the notation: "In
#13 (Stokols), pp. 169-189," referring the reader to the
appropriate pages in the volume PERSPECTIVES ON ENVIRONMENT
AND BEHAVIOR (D. Stokols, ed.).

     c. Each chapter starts with a brief commentary on the
area with which the chapter deals, and on the nature of the
literature on that topic, followed by the listing of items
relating to the topic, organized into appropriate sections.
Within each subsection, the material is listed in the
following sequence: 1   General books and monographs
(including readers limited to the topic of the chapter); 2.
journal articles and chapters from edited books, and reports
of governmental agencies or academic institutions; 3.
dissertations. Within each of these categories, material is
listed alphabetically.

     d. Wherever appropriate, items included under one
heading have been cross-referenced under other relevant
headings. This has been done by appending to the end of
each section a set of item numbers referring to such items,
with the notation: "See also: citations."

     In conclusion, let us call attention to a volume that
complements ours as a basic reference for the Environment-
and-Behavior field. It is item #7 of our Bibliography,
RESOURCES IN ENVIRONMENT AND BEHAVIOR (edited by W. White).
It provides basic information (unfortunately already out-of-
date in certain areas) about major centers for research and
graduate training, journals, funding agencies, and persons
identified with the field, as well as a useful list of major
monographs, annotated at much greater length than those in
our own volume.

Part I

GENERAL MATERIALS

1. GENERAL REFERENCE AND BIBLIOGRAPHIES

a. Bibliographies, Literature Reviews, and Reference Books

1. Bell, G., & MacGreevey, P. Behavior and environment: A bibliography of social activities in urban space. Monticello, Ill.: Council of Planning Librarians, 1970. (Exchange Bibliographies, No. 123).

   180+ items, annotated, covering diverse aspects of behavior relevant to urban life. Includes author and subject index.

2. Craik, K. H. Environmental psychology. Annual Review of Psychology 1973, 24, 402-422.

   A descriptive review of the environmental psychology literature through mid-1972.

3. Harrison, J. D. The perception and cognition of environment. Monticello, Ill.: Council of Planning Librarians, 1974. (Exchange Bibliographies, No. 516) 78 pp.

   Includes such topics as personality and perception, attitude formation, behavioral manipulation, and measurement and scaling. There are 690 references, one-fourth of which are annotated.

4. Morrison, D. E., Hornback, K. E., & Warner, K. W. ENVIRONMENT: A BIBLIOGRAPHY OF SOCIAL SCIENCE AND RELATED LITERATURE. Washington, D. C.: Office of Research and Monitoring, U.S. Environmental Protection Agency, 1973. 860 pp.

An unannotated listing of 4892 items covering
environmental psychology, sociology and other environmental
social sciences and related professional fields. In a final
285-page section, the titles of the items are indexed under
42 topical headings, arranged within each in the order of
their numbering in the bibliography.

5.   Pastalan, L. (Ed.). MAN-ENVIRONMENT REFERENCE:
          ENVIRONMENTAL ABSTRACTS. Ann Arbor, Michigan:
          Architectural Research Laboratory, University of
          Michigan, 1974. 418 pp.

A reference volume containing detailed 500-1000 word
synopses of 424 articles and studies. A composite index
allows accession of items in terms of population,
environment studied, and behavioral outcome.

6.   Stokols, D. Environmental psychology. Annual Review of
          Psychology, 1978, 29, 253-295.

An integrative review of the literature in the field,
divided into four categories: interpretive (cognitive
representation, personality and environment); evaluative
(environmental attitudes and assessment); operative
(behavioral analysis of ecologically relevant responses;
proxemic behavior), and responsive (environmental impact;
ecological psychology).

7.   White, W. (Ed.), RESOURCES IN ENVIRONMENT AND BEHAVIOR.
          Washington, D. C.: American Psychological
          Association, 1979. xi, 376 pp.

A general reference work covering the field of
environment and behavior, including a historical overview, a
listing of graduate programs, a section on teaching
innovations, an annotated bibliography of books in the
field, a chapter on career opportunities, sources of funding

for research, a list of journals, a list of organizations, and a personal-interest inventory of persons in the field.

b. Texts, Readers and Major Monographs

This section is comprised of textbooks, collections of readings and special journal issues containing original contributions and/or reprinted material; major integrative or theoretical treatments of the field are also included. The first volumes listed are considered basic reference materials, among which are serials devoted to reviews of research and theory on major problems in the field. For that reason the entire contents of each volume is listed. Individual papers are found in their particular topical categories. The remaining listings follow this initial set in a single alphabetical order.

8. Altman, I., & Wohlwill, J. F. (Eds.). HUMAN BEHAVIOR AND ENVIRONMENT: ADVANCES IN THEORY AND RESEARCH, Vol. 1. New York: Plenum, 1976. xv, 301 pp. Contents:

O'Riordan, T. Attitudes, behavior, and environmental policy issues. Pp. 1-36.

Wohlwill, J. F. Environmental aesthetics: The environment as a source of affect. Pp. 37-86.

Zube, E. H. Perception of landscape and land use. Pp. 87-122.

Mercer, D. C. Motivational and social aspects of recreational behavior. Pp. 123-162.

Parsons, H. M. Work environments. Pp. 163-210.

Willems, E. P. Behavioral ecology, health status, and health care: Applications to the rehabilitation setting. Pp. 211-264.

Schooler, K. K. Environmental change and the elderly. Pp. 265-298.

9. Altman, I., & Wohlwill, J. F. (Eds.). HUMAN BEHAVIOR AND
        ENVIRONMENT: ADVANCES IN THEORY AND RESEARCH,
        Vol. 2. New York: Plenum, 1977, xvi, 341 pp.
        Contents:

    Golledge, R. F. Multidimensional analysis in the
    study of environmental behavior and environmental
    design. Pp. 2-46.

    Appleyard, D. Understanding professional media:
    Issues, theory, and a research agenda. Pp. 47-88.

    Lazarus, R. S., & Cohen, J. B. Environmental
    stress. Pp. 89-128.

    Cone, J. D., & Hayes, S. C. Applied behavior
    analysis and the solution of environmental
    problems. Pp. 129-180.

    Altman, I., & Vinsel, A. M. Personal space: An
    analysis of E. T. Hall's proxemics framework. Pp.
    181-260.

    Klausner, S. Z. Energy and the structuring of
    society: Methodological issues. Pp. 261-306.

    Mann, S. H. The use of social indicators in
    environmental planning. Pp. 307-330.

10. Altman, I., & Wohlwill, J. F. (Eds.). HUMAN BEHAVIOR
        AND ENVIRONMENT, Vol. 3: CHILDREN AND THE
        ENVIRONMENT, New York: Plenum, 1978. xv, 300 pp.
        Contents:

    Tuan, Y. Children and the natural environment.
    Pp. 5-32.

    Parke, R. D. Children's home environments:
    Social and cognitive effects. Pp. 33-82.

    Moore, R., & Young, D. Childhood outdoors:
    Toward a social ecology of the landscape. Pp.
    83-130.

Gump, P. V. School environments. Pp. 131-174.

Wolfe, M. Childhood and privacy. Pp. 175-222.

Siegel, A. W., Kirasic, K. C., & Kail, R. V., Jr.
Stalking the elusive cognitive map: The
development of children's representations of
geographic space. Pp. 223-258.

Nagy, J. N., & Baird, J. C. Children as
environmental planners. Pp. 259-293.

11. Altman, I., Wohlwill, J. F., & Rapoport, A. (Eds.).
HUMAN BEHAVIOR AND ENVIRONMENT, Vol. 4:
ENVIRONMENT AND CULTURE. New York: Plenum, 1980.
Contents:

Rapoport, A. Cross-cultural aspects of
environmental design. Pp. 7-46.

Brislin, R. W. Cross-cultural research methods:
Strategies, problems, applications. Pp. 47-82

Berry, J. W. Cultural ecology and individual
behavior. Pp. 83-106

Aiello, J. R., & Thomason D. Personal space,
crowding, and spatial behavior in a cultural
context. Pp. 107-178.

Brower, S. Territory in urban settings. Pp.
179-208.

Richardson, M. Culture and the urban stage: The
nexus of setting, behavior and image in urban
places. Pp. 209-242.

Bennett, J. W. Human ecology as human behavior: A
normative anthropology of resource use and abuse.
Pp. 243-277.

Sorensen, J. & White, G. F. Natural hazards: A
cross-cultural perspective. Pp. 279-318.

Sachs, I. Culture, ecology and development:
Redefining planning approaches. Pp. 319-343.

12. Baum, A., Singer, J. E., & Valins, S. (Eds.). ADVANCES
IN ENVIRONMENTAL PSYCHOLOGY. Vol. I: THE URBAN
ENVIRONMENT. Hillsdale, N. J.: L. Erlbaum, 1978.
xiii, 204 pp. Contents:

Cohen, S. Environmental load and the allocation
of attention. Pp. 1-30.

Milgram, S., & Sabini, J. On maintaining urban
norms: A field experiment in the subway. Pp.
31-40.

Singer, J. E., Lundberg, U., & Frankenhaeuser, M.
Stress on the train: A study of urban commuting.
Pp. 41-56.

Baron, R. A. Agression and heat: The "long hot
summer" revisited. Pp. 57-84.

Korte, C. Helpfulness in the urban environment.
Pp. 85-110.

Stokols, D. In defense of the crowding construct.
Pp. 111-130.

Fischer, C. S. Sociological comments on
psychological approaches to urban life. Pp.
131-144.

Baron, R. M., & Rodin, J. Personal control as a
mediator of crowding. Pp. 145-192.

13. Stokols, D. (Ed.). PERSPECTIVES ON ENVIRONMENT AND
BEHAVIOR. New York: Plenum, 1976. xiv, 360 pp.
Contents:

Stokols, D. Origins and directions of
environment-behavioral research. Pp. 5-36.

Willems, E. P. Behavioral ecology. Pp. 39-68.

Wicker, A. W., & Kirmeyer, S. From church to laboratory to national park: A program of research on excess and insufficient populations in behavior settings. Pp. 69-96.

Proshansky, H. M., & O'Hanlon, T. Environmental psychology: Origins and development. Pp. 101-130.

Glass, D. C., Singer, J. E., & Pennebaker, J. W. Behavioral and physiological effects of uncontrollable environmental events. Pp. 131-152.

Loo, C. Beyond the effects of crowding: Situational and individual differences. Pp. 153-168.

McKechnie, G. E. Simulation techniques in environmental psychology. Pp. 169-190.

Sommer, R. Action research. Pp. 195-204.

Michelson, W. From congruence to antecedent conditions: A search for the basis of environmental improvement. Pp. 205-220.

Kaplan, S. Participation in the design process: A cognitive approach. Pp. 221-234.

Kaplan, R. Preference and everyday nature: Method and application. Pp. 235-250.

Kravetz, M. L. Who needs what when: Design of pluralistic learning environments. Pp. 251-272.

Baldassare, M., & Fischer, C. S. The relevance of crowding experiments to urban studies. Pp. 273-286.

Smith, M. B. Some problems of strategy in environmental psychology. Pp. 289-302.

Altman, I. Research on environment and behavior: A personal statement of strategy. Pp. 303-324.

Patterson, A. H., Methodological developments in
environment-behavioral research. Pp. 325-344.

14. Wohlwill, J. F., & Carson, D. H. (Eds.). ENVIRONMENT
    AND THE SOCIAL SCIENCES: PERSPECTIVES AND
    APPLICATIONS. Washington, D. C.: American
    Psychological Association, 1972. 300 pp.
    Contents:

    Marsden, H. M. Crowding and animal behavior. Pp.
    5-14.

    Esser, A. H. A biosocial perspective on crowding.
    Pp. 15-28.

    Proshansky, H. M., Ittelson, W. H., & Rivlin, L.
    Freedom of choice and behavior in a physical
    setting. Pp. 29-43.

    Zlutnick, S., & Altman, I. Crowding and human
    behavior. Pp. 44-60.

    Jones, M. H. Pain thresholds for smog components.
    Pp. 61-65.

    Swan, J. A. Public response to air pollution.
    Pp. 66-74.

    Buckhout, R. Pollution and the psychologist: A
    call to action. Pp. 75-86.

    Agron, G. Some observations on behavior in
    institutional settings. Pp. 87-94.

    Ittelson, W. H., Proshansky, H. M., & Rivlin, L.
    Bedroom size and social interaction of the
    psychiatric ward. Pp. 95-104.

    Glaser, D. Architectural factors in isolation
    promotion in prisons. Pp. 105-113.

    Lawton, M. P. Some beginnings of an ecological
    psychology of old age. Pp. 114-125.

Yancey, W. L. Architecture, interaction, and social control: The case of a large-scale housing project. Pp. 126-136.

Fried, M., & Gleicher, P. Some sources of residential satisfaction in an urban slum. Pp. 137-153.

Carson, D. H. Residential descriptions and urban threats. Pp. 154-168.

Zehner, R. B. Neighborhood and community satisfaction: A report on new towns and less planned suburbs. Pp. 169-183.

Burton, I. Cultural and personality variables in the perception of natural hazards. Pp. 184-197.

Lime, D. W. Behavioral research in outdoor recreation management: An example of how visitors select campgrounds. Pp. 198-206.

Schafer, E. L., Jr., & Mietz, J. Aesthetic and emotional experiences rate high with northeast wilderness hikers. Pp. 207-216.

Marans, R. W. Outdoor recreation behavior in residential environments. Pp. 217-232.

Driver, B. L. Potential contributions of psychology to recreation resource management. Pp. 233-248.

Sewell, W. R. D. Environmental perceptions and attitudes of engineers and public health officials. Pp. 249-264.

Driver, B. L. Professional socialization and environmental management. Pp. 265-278.

Studer, R. G. The organization of spatial stimuli. Pp. 279-292.

Epilogue. Environment and behavioral science: Retrospect and prospect. Pp. 293-300.

15. Bell, P. A., Loomis, R., & Fisher, J. D. ENVIRONMENTAL
      PSYCHOLOGY. Philadelphia: Saunders, 1978. xivx,
      457 pp.

    A text, providing comprehensive coverage of the field.
Includes 27 pages of references.

16. Canter, D. PSYCHOLOGY FOR ARCHITECTS. London:
      Applied Science Publishers, 1974. xi, 171 pp.

    A brief treatment of aspects of psychology relevant to
architecture. Includes chapters on perception, learning and
development, individual differences, dimensions of
environmental meaning, and use of space.

17. Canter, D., & Stringer, P. (Eds.). ENVIRONMENTAL
      INTERACTION. New York: International
      Universities Press, 1975. 374 pp.

    An introductory environmental psychology text with
chapters contributed by several authors. Discussed are
thermal, acoustic, luminous and spatial environments,
building usage, understanding and living in the city, the
natural environment, and contributions made in environmental
decision-making.

18. Friedman, S., & Juhasz, J. B. (Eds.). ENVIRONMENTS:
      NOTES AND SELECTIONS ON OBJECTS, SPACES, AND
      BEHAVIOR. Monterey, California: Brooks-Cole,
      1974. 275 pp.

    An assorted set of readings taken from diverse sources,
including psychology, sociology, architecture and the arts.

19. Heimstra, N. W., & McFarling, L. H. ENVIRONMENTAL
      PSYCHOLOGY (2nd ed.). Monterey, California:
      Brooks-Cole, 1978. 277 pp.

A general text, at an elementary level.

20. Holahan, C. J. ENVIRONMENT AND BEHAVIOR: A DYNAMIC
     PERSPECTIVE. New York: Plenum, 1978. xvi, 188
     pp.

Intended as a text, this book consists of a series of
essays, based in part on reports of the author's research,
on the coping process in the individual's response to the
built environment, and the relationship between coping and
satisfaction with settings such as hospitals, dormitories,
etc.

21. Ittelson, W. H., Proshansky, H. M., Rivlin, L. G., &
     Winkel, G. H. AN INTRODUCTION TO ENVIRONMENTAL
     PSYCHOLOGY. New York: Holt, Rinehart, and
     Winston, 1974. 406 pp.

A comprehensive text, containing chapters on theory and
methods, including coverage of both the man-made and the
natural environment.

22. Kaplan, S., & Kaplan, R. (Eds.), HUMANSCAPE:
     ENVIRONMENTS FOR PEOPLE. North Scituate, Mass.:
     Duxbury Press, 1978. xv, 480 pp.

A reader for undergraduate courses in environmental
psychology and related subjects. Includes material from
both scientific and humanistic literature, as well as
magazines. Several original selections are also included.

23. Kates, R. W., & Wohlwill, J. F. (Eds.). Man's response
     to the physical environment. Journal of Social
     Issues, 1966, 22 (4), 1-140.

An early collection of contributions to the environment-
behavior field from geography, psychology, sociology and
design, including considerations of environmental
adaptation, use of space, site planning, and architectural
programming.

24. Lee, T. R. PSYCHOLOGY AND THE ENVIRONMENT. London:
        Methuen, 1976. 143 pp.

A paperback providing a brief introduction to some
aspects of environmental psychology, focusing particularly
on the social environment, including residential,
institutional and urban environments.

25. Leff, H. L. EXPERIENCE, ENVIRONMENT, AND HUMAN
        POTENTIALS. New York: Oxford University Press,
        1978. 523 pp.

An integrative treatment of the interplay of cognitive
and motivational forces in the individual's response to the
environment, with attention given to environmental
cognition, environmental attitude and values, the
development of ecological consciousness, and the role of
design and environmental planning in optimizing human
satisfaction.

26. Moos, R. H. THE HUMAN CONTEXT: ENVIRONMENTAL
        DETERMINANTS OF BEHAVIOR. New York: Wiley, 1976.
        xi, 444 pp.

A broad treatment of the role of environmental
influences on behavior, including a wide range of factors
such as weather, noise, and air pollution, the architectural
environment, organizational structure, and social climate.

27. Moos, R. H., & Insel, P. M. (Eds.). ISSUES IN SOCIAL
        ECOLOGY: HUMAN MILIEUS. Palo Alto, California:
        National Press, 1974. 616 pp.

A collection of previously published material, covering physical, social, organizational and institutional aspects of environment-behavior relationships.

28. Porteous, J. D. ENVIRONMENT AND BEHAVIOR: PLANNING AND
        EVERYDAY URBAN LIFE. Reading, Massachusetts:
        Addison-Wesley, 1977. 446 pp.

A text on environment-behavior relations, emphasizing urban-environmental settings, and directed at planning students.

29. Proshansky, H. M., Ittelson, W. H., & Rivlin, L. G.
        (Eds.). ENVIRONMENTAL PSYCHOLOGY: MAN AND HIS
        PHYSICAL SETTING. New York: Holt, Rinehart, and
        Winston, 1970. 690 pp.

An anthology of papers largely reprinted from other sources. Topics range from theory and basic processes to social institutions, and design and planning.

30. Proshansky, H. M., Ittelson, W. H., & Rivlin, L. G.
        (Eds.). ENVIRONMENTAL PSYCHOLOGY; PEOPLE AND
        THEIR PHYSICAL SETTINGS (2ND ED.). New York: Holt,
        Rinehart, and Winston, 1976. 632 pp.

A second anthology of papers, mostly reprinted from other sources, overlapping only minimally with the selections chosen for the first edition, and offering a sampling of predominantly theoretical and essay-type articles on diverse facets of the field.

31. Psychology and ecology: Environmental quality and
        human adjustment. International Journal of
        Psychology, 1977,12, 69-157.

Includes general theoretical and methodological articles as well as material on quality of life, population psychology, problems of industrial work, etc.

32. Saarinen, T. F. ENVIRONMENTAL PLANNING: PERCEPTION AND BEHAVIOR. Boston: Houghton-Mifflin, 1976. xiii, 262 pp.

An interesting overview of the environment-behavior field by a behavioral geographer. Each of the chapters presents phenomena at a particular scale, ranging from a room to the world.

33. Sims, J. H., & Baumann, D. D. (Eds.). HUMAN BEHAVIOR AND ENVIRONMENT: INTERACTIONS BETWEEN MAN AND HIS PHYSICAL WORLD: Chicago: Maaroufa, 1974. 354 pp.

A highly diverse set of reprinted readings culled from psychology, psychiatry, geography, anthropology, and sociology.

c. Conference Proceedings

Listed are the proceedings of the major conference series in the environment-behavior field such as the Architectural Psychology Series from Britain, and the series sponsored by the Environmental Design Research Association in the U.S. (EDRA). Proceedings of other Conferences, held under diverse auspices and occurring at irregular intervals or only once, are listed following the EDRA volumes, arranged alphabetically by senior editors' names.

EDRA Conference Proceedings

Since 1969, the Environmental Design Research Association, EDRA, has held yearly conferences on different university campuses in the U.S. and Canada. The published

proceedings from these conferences cover diverse fields,
disciplines, and material, including reports of original
research, theoretical and methodological papers, discussions
of professional and applied issues, and abstracts of
workshops. Because of the very large volume of material
contained in these publications, its unedited nature and
rather variable quality, it was not practicable to itemize
their contents. Selected individual papers from these
volumes are, however, included in their respective topical
sections.

34. Sanoff, H. & Cohn, S. (Eds.) EDRA ONE: PROCEEDINGS OF
        THE FIRST ANNUAL ENVIRONMENTAL DESIGN RESEACH
        ASSOCIATION CONFERENCE. Raleigh, North Carolina:
        North Carolina State University, 1969.
        (Republished: Stroudsburg, Pa.: Dowden,
        Hutchinson, and Ross. 1975). 384 pp.

35. Archea, J., & Eastman, C. (Eds.). EDRA TWO:
        PROCEEDINGS OF THE SECOND ANNUAL ENVIRONMENTAL
        DESIGN RESEARCH ASSOCIATION CONFERENCE.
        Pittsburgh, Pa.: Carnegie Mellon University, 1970
        (Republished: Stroudsburg, Pa.: Dowden,
        Hutchinson, and Ross, 1975). 408 pp.

36. Mitchell, W. J. (Ed.). ENVIRONMENTAL DESIGN: RESEARCH
        AND PRACTIC (Vols. 1 & 2). Proceedings of the
        EDRA 3 Conference. Los Angeles: University of
        California, 1972.

37. Preiser, W. F. E. (Ed.). ENVIRONMENTAL DESIGN
        RESEARCH. (Vol. I: SELECTED PAPERS; Vol. II:
        SYMPOSIA AND WORKSHOPS) Proceedings of the Fourth
        Environmental Design Research Association
        Conference, Virginia State University and
        Polytechnic Institute, Blacksburg, Va., 1973.
        Stroudsburg, Pa.: Dowden, Hutchinson, and Ross,
        1973. 568, 560 pp.

38.   Carson, D. H. (Ed.). MAN-ENVIRONMENT INTERACTIONS:
        EVALUATION AND APPLICATIONS (Vols. 1-12).
        Proceedings of the Fifth Environmental Design
        Research Association Conference, University of
        Wisconsin, Milwaukee, 1974. Washington, D C.:
        Environmental Design Research Association, 1974.
        1524 pp.

        The conference divided its proceedings among the
        following volumes:

        Vol. 1. Carson, D. H. (Ed.). MAN-ENVIRONMENT
        THEMES.

        Vol. 2. Wolf, C. P. (Ed.). SOCIAL IMPACT
        ASSESSMENT.

        Vol. 3. Parsons, H. M. (Ed.). HUMAN FACTORS.

        Vol. 4. Davis, G. (Ed.). FIELD APPLICATIONS.

        Vol. 5. Lozar, C. C. (Ed.). METHODS AND
        MEASURES.

        Vol. 6. Margulis, S. T. (Ed.). PRIVACY.

        Vol. 7. Chase, R. A. (Ed.). SOCIAL ECOLOGY.

        Vol. 8. Bechtel, R. B. (Ed.). UNDERMANNING
        THEORY.

        Vol. 9. Blasdel, H. G. (Ed.). MULTIVARIATE
        METHODS.

        Vol. 10. Bazjanec, V. (Ed.). COMPUTERS AND
        ARCHITECTURE.

        Vol. 11. Honikman, B. (Ed.). COGNITION AND
        PERCEPTION.

        Vol. 12. Moore, R. C. (Ed.). CHILDHOOD CITY.

39.   Honikman, B. (Ed.). RESPONDING TO SOCIAL CHANGE.
        Companion to the Sixth International Environmental

Design Research Association Conference, University
of Kansas, Lawrence, Ks. 1975. Stroudsburg, Pa.:
Dowden, Hutchinson, and Ross, 1975. 311 pp.

40. Suedfeld, P., & Russell, J. A. (Eds.). THE BEHAVIORAL
    BASIS OF DESIGN (Vols. 1-2). Proceedings of the
    Seventh Environmental Design Research Association
    Conference, University of British Columbia,
    Vancouver, Canada, 1976. Stroudsburg, Pa.:
    Dowden, Hutchinson, and Ross, 1977.

    Vol. I. Ward, L. M., Coren, S., Gruft, A., &
    Collins, J. B. (Eds.). SELECTED PAPERS. 381 pp.

    Vol. II. Suedfeld, P., Russell, J. A., Ward, L.
    M., Szigeti, F., & Davis, G. (Eds.). INVITED
    ADDRESSES, SYMPOSIA, AND WORKSHOPS.

41. Weideman, S., Anderson, J., & Brauer, R. (Eds.).
    PRIORITIES FOR ENVIRONMENTAL DESIGN RESEARCH.
    Proceedings of the Eighth Annual Conference of the
    Environmental Design Research Association,
    University of Illinois, Champaign-Urbana, Ill.,
    1977. Washington, D. C.: Environmental Design
    Research Association, 1978. 535 pp.

42. Rogers, W., & Ittelson, W. (Eds.). NEW DIRECTIONS IN
    ENVIRONMENTAL DESIGN RESEARCH. Proceedings of the
    Ninth Annual Conference of the Environmental
    Design Research Association, University of
    Arizona, Tucson, 1978. Washington, D. C.:
    Environmental Design Research Association, 1978.
    518 pp.

43. Seidel, A., & Danford, S. (Eds.). ENVIRONMENTAL
    DESIGN: RESEARCH, THEORY AND APPLICATION.
    Proceedings of the Tenth Annual Conference of the
    Environmental Design Research Association, State
    University of New York at Buffalo, Buffalo, N. Y.,
    1979. Washington, D.C.: Environmental Design
    Research Association, 1979. 448 pp.

44. Stough, R. R., & Wandersman, A. OPTIMIZING
        ENVIRONMENTS: RESEARCH, PRACTICE, AND POLICY.
        Proceedings of the Eleventh Annual Conference of
        the Environmental Design Research Association,
        College of Charleston, Charleston, S. C., 1980.
        Washington, D. C.: Environmental Design Research
        Association, 1980, 339 pp.

45. Bailey, R., Branch, C. H., & Taylor, C. W. (Eds.).
        ARCHITECTURAL PSYCHOLOGY AND PSYCHIATRY: AN
        EXPLORATORY NATIONAL RESEARCH CONFERENCE. Salt
        Lake City, Utah: University of Utah, 1961.

46. Canter, D. (Ed.). ARCHITECTURAL PSYCHOLOGY:
        PROCEEDINGS OF THE CONFERENCE AT DALANDHUI,
        UNIVERSITY OF STRATHCLYDE. London: Royal
        Institute of British Architects, 1969. 92 pp.

47. Honikman, B. (Ed.). AP 70: PROCEEDINGS OF THE
        ARCHITECTURAL PSYCHOLOGY CONFERENCE AT KINGSTON
        POLYTECHNIC 1970. London: Royal Institute of
        British Architects, 1971. 98 pp.

48. Canter, D., & Lee, T. PSYCHOLOGY AND THE BUILT
        ENVIRONMENT. New York: Wiley, 1974. 213 pp.

A set of papers from the Third British Conference on
Architectural Psychology (1974).

49. Korosec-Sarfaty, P. (Ed.). THE APPROPRIATION OF SPACE.
        Proceeding of the Third International
        Environmental Psychology Conference, Louis Pasteur
        University. Strasbourg: Strasbourg University
        Press, 1976.

50. Kuller, R., (Ed.). ARCHITECTURAL PSYCHOLOGY:
    PROCEEDINGS OF THE LUND CONFERENCE. Stroudsburg,
    Pa.: Dowden, Hutchinson, and Ross, 1974. 452 pp.

51. Taylor, C. W., Bailey, R., & Branch, C. H. H. (Eds.).
    NATIONAL CONFERENCE ON ARCHITECTURAL PSYCHOLOGY,
    2nd. Salt Lake City, Utah: University of Utah,
    1967.

2. HISTORICAL, THEORETICAL AND METHODOLOGICAL WORKS

a. History, Theory and Selective Reviews

This section includes presentations of theoretical positions and models, historical accounts of the field, discussions of its relations to psychology as a whole as well as to other disciplines, and selective, integrative reviews of literature.

52. Kruse, L. RAUMLICHE UMWELT: EINE PHAENOMENOLOGIE DES RAUMLICHEN VERHALTENS ALS BEITRAG ZU EINER PSYCHOLOGISCHEN UMWELTTHEORIE. Berlin: W. de Gruyter, 1974. 174 pp.

Translation of title: A phenomenology of spatial behavior, as a contribution to a theory of the psychological environment. (Not available for abstracting.)

53. Mehrabian, A., & Russell, J. A. AN APPROACH TO ENVIRONMENTAL PSYCHOLOGY. Cambridge, Mass.: MIT Press, 1974. 266 pp.

Statement of a theory and program of research, emphasizing the emotional impact of the environment, and the arousal properties of environmental stimulation.

54. Wapner, S., Cohen, S., & Kaplan, B. (Eds.). EXPERIENCING THE ENVIRONMENT. New York: Plenum, 1976. 244 pp.

A set of papers based on a conference presenting
different perspectives of environmental experience.
Includes phenomenological and organismic views (Ittelson et
al.; Kaplan et al.), the cultural-historical approach
(Lowenthal & Prince), arousal and stimulation theory
(Russell & Mehrabian; Wohlwill & Kohn), the ecological
(Wicker & Kirmeyer), and the personality paradigm (Craik;
Little). An additional paper (Kates) deals with the
environment as a hazard.

55. Windley, P. G., Byerts, T. O., & Ernst, F. G. (Eds.).
      THEORY DEVELOPMENT IN ENVIRONMENT AND AGING.
      Washington, D.C.: Gerontological Society, 1975.
      294 pp.

Proceedings of a conference directed at stimulating
research on the environmental aspects of problems of the
aged.

56. Altman, I.   Some perspectives on the study of man-
      environment phenomena. Representative Research in
      Social Psychology, 1973, 4, 109-126.

Compares the behavioral scientist and the design
practitioner on their differing orientations to man-
environment issues and their "models-of-man".

57. Altman,. I. Environmental psychology and social
      psychology. Personality and Social Psychology
      Bulletin, 1976, 2, 96-113.

An examination of environmental psychology and its
relationships to social psychology. Certain characteristics
of the former, including its molar perspective, problem-
oriented character and conceptual and methodological breadth
are stressed. Possibilities for a symbiotic relationship
between environmental and social psychology are noted.

58. Borden, R. J. One more look at social and
         environmental psychology: Away from the looking
         glass and into the future. Personality and Social
         Psychology Bulletin, 1977, 3, 407-411.

    Reviews previous controversy over the status of
environmental psychology, and suggests the area of
environmental concern as an emerging point of convergence
between environmental and social psychology.

59. Chein, I. The environment as a determinant of
         behavior. Journal of Social Psychology 1954, 39,
         115-127.

    A theoretical model of the "geo-behavioral" environment
in terms of four features: stimuli, goal objects, supports
and directors.

60. Craik, K. H. The comprehension of the everyday
         physical environment. Journal of the American
         Institute of Planners, 1968, 34, 29-37.

    A brief overview of approaches and methodological issues
in the study of people's response to environmental
information, with particular emphasis on modes of presenting
such information and on the response formats used to
investigate correlated behaviors.

61. Craik, K. H. Environmental psychology. In NEW
         DIRECTIONS IN PSYCHOLOGY (Vol. 4). New York:
         Holt, Rinehart & Winston, 1970. Pp. 1-122.

    A comprehensive overview of early work in environmental
psychology and related topics, with special attention to the
built environment, natural hazards, and environmental
attitudes and values.

62. Craik, K. H. Multiple scientific paradigms in
         environmental psychology: International Journal
         of Psychology, 1977, 12, 147-157.

    Major paradigms of environmental-psychology research are
illustrated with reference to problems of ecological
psychology, environmental perception, cognition and
assessment, personality and the environment, and functional
adaptation to the environment.

63. Epstein, Y. M. Comment on environmental psychology and
         social psychology. Personality and Social
         Psychology Bulletin, 1976, 2, 346-349.

    A comment on Altman's paper (#57), delineating
relationships between environmental and social psychology.
Altman's view is thought to characterize the broad field of
"environment-and-behavior" and the work of non-
psychologists, along with some non-social environmental
psychologists, rather than social psychologists working on
problems of crowding and personal space.

64. Ittelson, W. H. Some issues facing a theory of
         environment and behavior. In #30, (Proshansky et
         al.), 1976, pp. 51-59.

    A discussion of theoretical and epistemological issues
in the environment-and-behavior field, stressing the need to
overcome the dichotomization of the world into environment
and person, and focusing on the study of choice, action and
change.

65. Kahana, E. A congruence model of person-environment
         interaction. In #55. (Windley et al.), 1975, pp.
         181-214.

    An analysis of man-environment interactions in terms of
the congruence between the environment and the needs of the

individual. Seven dimensions of congruence are
differentiated: segregate, congregate, institutional
control, structure, stimulation, affect, and impulse
control.

66. Kaplan, S. The challenge of environmental psychology:
    A proposal for a new functionalism. American
    Psychologist, 1972, 27, 140-143.

   A plea for a functional approach to the study of
behavior in relation to both environmental problems and
human response to environmental conditions.

67. Kates, R. W. Human perception of the environment.
    Social Science Journal, 1970, 22, 648-660.

   A brief general review focusing on four topics:
illusory perception, urban images, environmental attitudes
in relation to landscape, and adjustment to drought.

68. Kaye, S.M. Psychology in relation to design: An
    overview. Canadian Psychological Review, 1975,
    16, 104-110.

   A brief overview of environmental-psychology research of
relevance to design.

69. Krupat, E. Environmental and social psychology: How
    different must they be? Personality and Social
    Psychology Bulletin, 1977, 3, 51-53.

   Argues that environmental psychology represents an
outward extension of the perspective of psychology from the
individual to the social environment and finally to the
physical environment.

70. Moos, R. H. Conceptualizations of human environments.
         American Psychologist, 1973, 28, 652-665.

     Human environments are considered in terms encompassing
geographic, ecological and physical aspects, as well as
social, institutional and organizational ones.

71. Pervin, L. A. Definitions, measurements, and
         classifications of stimuli, situations, and
         environments. Human Ecology, 1978, 6, 71-105.

     A review of definitional and taxonomic issues, with
emphasis on molar vs. molecular levels of analysis,
objective vs. subjective definition of stimuli, and
formulation of the relationship between environmental and
behavioral variables.

72. Peterson R., Wekerle, G. R., & Morley, D. Women and
         environments: An overview of an emerging field.
         Environment and Behavior, 1978, 10, 511-534.

     An overview of issues concerning the differing responses
to the built and natural environments by women and men.
Includes a discussion of specific environments for women.

73. Proshansky, H. M. Comment on environmental and social
         psychology. Personality and Social Psychology
         Bulletin, 1976, 2, 359-363.

     Questions the prospects of a symbiotic relationship
between social and environmental psychology suggested by
Altman (#57) in view of the dominant mode of research and
thinking in social psychology, which is considered to be
unsuited to the needs of a "real-world" oriented
environmental psychology.

74. Proshansky, H. M. Environmental psychology and the
        real world. American Psychologist, 1976, 31,
        303-310.

    A personal statement presenting the author's conception
of environmental psychology as a field separate from the
realm of social psychology, particularly in its focus on
real-life problems. Five features characterizing
environmental psychology are listed, among them an
orientation towards both content and context, and a focus on
the temporal dimension in the analysis of problems.

75. Proshansky, H. M., & O'Hanlon, T. Environmental
        psychology: Origins and development. In #13
        (Stokols), pp. 101-129.

    An analysis of the history and current status of the
field.

76. Rapoport, A. An approach to the construction of man-
        environment theory. In #37 (EDRA-4, Vol. II), pp.
        124-135.

    An examination of the role of theory, and of the
different kinds of theories and models, in man-environment
studies.

77. Sells, S. B. Dimensions of stimulus situations which
        account for behavior variance. In S. B. Sells
        (Ed.), STIMULUS DETERMINANTS OF BEHAVIOR. New
        York: Ronald, 1963. Pp. 3-15.

    An attempt at the development of a taxonomy of
environmental, situational and task-related variables that
affect behavior, including both physical and social factors.

78.   Sonnenfeld, J.   Variable values in space and landscape:
         An inquiry into the nature of environmental
         necessity.   Journal of Social Issues, 1966, 22
         (4), 71-82.

     A discussion of the differing expectations of natives
and non-natives with regard to the availability of open
space in residential areas.   Makes a critical distinction
between two contrasting modes of responding to experienced
environmental mis-match, or dissatisfaction: Adaptation
(change in the individual's response to the environment)
versus Adjustment (change of the environment by the
individual).

79.   Stokols, D.   Social unit analysis as a framework for
         research in environmental and social psychology.
         Personality and Social Psychology Bulletin, 1976,
         2, 350-358.

     Altman's concept of "social-unit analysis" is examined
in relation to issues of goal-optimization, and in terms of
its place in environmental psychology.

80.   Stokols, D.   Origins and directions of environment-
         behavioral research.   In #13 (Stokols), pp. 5-36.

     An overview of the field of environment-and-behavior,
comparing the approaches of ecological and environmental
psychology and pointing to future integrative development of
the field at the levels of theory and research.

81.   Studer, R. G.   Man-environment relations: Discovery or
         design?   In #37, (EDRA-4, Vol. II), pp. 136-151.

     The relationship between environment-behavior research
and environmental design is examined in terms of the place
accorded to intervention in a system.   Environmental
management, through the application of behavior technology,

is contrasted to noninterventive approaches such as
behavioral ecology.

82. Tibbets, P. Epistemology, perceptual theory, and the
        built environment. Man-Environment Systems, 1976,
        6, 91-98.

    A discussion of certain theoretical and epistemological
issues from the domain of the psychology of perception
considered to have relevance for conceptualizations of
response to the built environment. Four epistemological
positions are considered: naive realism, representative
realism, phenomenalism, and transactionalism.

83. Tuan, Yi-Fu. Structuralism, existentialism, and
        environmental perception. Environment and
        Behavior, 1972, 4, 319-331.

    A discussion of the potential contribution of the
philosophical principles of structuralism and existentialism
to the understanding of environmental perception.

84. Wapner, S., Kaplan, B., & Cohen S. B. An organismic-
        developmental perspective for understanding
        transactions of men in environments. Environment
        and Behavior, 1973, 5, 255-289.

    Presents a perspective on environment-behavior problems
grounded in organismic-developmental theory, emphasizing the
role of individuals in structuring, interpreting, orienting
themselves in and experiencing their environment, and the
relationship between the environment as perceived and
construed and actions directed at the environment.

85. Wohlwill, J. F. The emerging discipline of
        environmental psychology. American Psychologist,
        1970, 25, 303-312.

A brief account of trends and developments relating to
the emergence of environmental psychology as a discipline.

86. Wohlwill, J. F. The environment is not in the head!
         In #37, (EDRA-4, Vol. II), pp. 166-181.

Argues for the desirability of an objective, physical-
environment based definition of environmental variables in
environment-behavior research, as opposed to current
approaches that start from the environment as perceived,
imaged or interpreted.

87. Wohlwill, J. F. Psychology and the environmental
         disciplines. In M. Bornstein (Ed.), PSYCHOLOGY
         AND ITS ALLIED DISCIPLINES. Hillsdale, N. J.: L.
         Erlbaum, Assoc. (In press.)

A discussion of the diverse links of psychology to other
fields concerned with environmental problems,
differentiating between relations to other social-science
disciplines, such as geography and sociology, and to
professional fields such as architecture and natural
recreation.

See also: citations 239, 587, 1436.

b. Methodology

88. Michelson, W. M. (Ed.) BEHAVIORAL RESEARCH METHODS IN
         ENVIRONMENTAL DESIGN. Stroudsburg, Pa.: Dowden,
         Hutchinson & Ross, 1975. 307 pp.

Includes chapters on research-design stratgegy, use of semantic differential and other paper-and-pencil techniques, trade-off games, survey research, time-budget analysis, photographic recording and observational methods.

89. Appleyard, D. Understanding professional media:
         Issues, techniques and research agenda. In #9
         (Altman & Wohlwill, 2), pp. 43-88.

An overview of approaches and techniques of environmental stimulation through diverse media and their use in planning.

90. Berglund, B. Quantitative approaches in environmental
         studies. International Journal of Psychology,
         1977, 12, 111-124.

A discussion of the application of methods of psychological scaling in the study of noise annoyance, temperature discomfort, response to odors and pollutants, as well as other problems of environmental health.

91. Berglund, B. et al. On the scaling of annoyance due to
         environmental factors. Environmental Psychology
         and Nonverbal Behavior, 1977, 2, 83-92.

Discusses problems of psychological scaling (within a Thurstonian framework) and unit calibration in the measurement of annoyance responses to odors and noises.

92. Craik, K. H. The assessment of places. In P. M.
         McReynolds (Ed.) ADVANCES IN PSYCHOLOGICAL
         ASSESSMENT (Vol. 2), Palo Alto, California.:
         Science and Behavior Books, 1971. Pp. 40-62.

Compares environmental assessment with personality
assessment, and reviews techniques available for the
assessment of places, modeled on those in use in personality
measurement.

93. Danford, S. & Willems, E. P. Subjective responses to
     architectural displays: A question of validity.
     Environment and Behavior, 1975, 7, 486-516.

Questions the interpretation of verbal ratings of
environmental stimuli, on the basis of a study in which
profiles and factor loadings were compared for ratings of a
building (a) in the field, (b) photographically presented,
or (c) verbally described; these results pointed to the role
of instrument bias and showed the importance of convergent
as well as divergent validation of methodologies.

94. Golledge, R. G. Multidimensional analysis in the study
     of environmental design. In #9 (Altman &
     Wohlwill, 2), pp. 1-42.

A presentation of the principles of multidimensional
scaling, and other multivariate techniques (e.g., factor and
cluster analysis), illustrated with examples from the area
of the cognition of urban space, and other environment-
behavior problems.

95. Kasmar, J. V. The development of a usable lexicon of
     environmental descriptors. Environment and
     Behavior, 1970, 2, 153-169.

Describes the procedure followed in devising the
Environment Description Scale, consisting of 66 bi-polar
adjective rating scales, and its application to the
description of rooms.

96. Knight, R. L. & Menchik, M. D. Conjoint preference
     estimation for residential land use policy

evaluation. In #131 (Golledge & Rushton), pp.
135-155.

An interesting application of conjoint measurement
procedures (a variant of multidimensional scaling) to
determine people's evaluations of trade-offs relating to
residential design (e.g., between views from a house and
backyard size).

97. May, H. B., & Basalla, T. G. Comparison of factor
         analysis and multidimensional scaling to analyze
         semantic differential and construct eliciting
         data. In #40 (EDRA-7, Vol. II), pp. 1-9.

A comparison of two analytic techniques to determine
images of an urban neighborhood obtained via semantic
differential ratings and repertory grid construct data.

98. McKechnie, G. E. The environmental response inventory
         in application. Environment and Behavior, 1977,
         9, 255-276.

A survey of research utilizing the author's
Environmental Response Inventory, an instrument designed to
measure people's dispositions towards different aspects and
dimensions of their environment. Applications include
studies of individual and group difference in migration,
environmental decision-making, and choice of leisure
activities.

99. McKechnie, G. E. Simulation techniques in
         environmental psychology. In #13 (Stokols), pp.
         169-189.

A discussion of different modes of simulation of
physical environments and the problems they raise. Includes
a description of the Berkeley Environmental Simulation
Laboratory and its application in research.

100. Mehrabian, A. & Russell, J. A. A verbal measure of
      information rate for studies in environmental
      psychology. Environment and Behavior, 1974, 6,
      233-252.

      Argues for the use of information rate as a concept of
high degree of generality applicable to research on arousal
produced by environmental stimulation and on affective
response to the environment. A verbal measure of
information rate is proposed to provide a readily usable
means of operationalizing this concept, and data on its
reliability and validity are presented.

101. Patterson, A. H. Unobtrusive measures: Their nature
      and utility for architects. In #1472 (Lang), pp.
      261-273.

      A presentation of approaches to studying behavior in
field settings based on observations and records correlated
with behavior that do not involve intrusion on the
individual's activity or awareness.

102. Patterson, A. H. Methodological developments in
      environment-behavioral research. In #13
      (Stokols), pp. 325-344.

      Discusses general methodological issues of reliability
and validity, research design and methods of data
collection, relevant to environment-behavior research.

103. Sanoff, H. Measuring attributes of the visual
      environment. In #1472 (Lang), pp. 244-260.

      A semantic-differential analysis of ratings of the
appearance of row-house street blocks viewed from
photographs.

104. Thiel, P. Notes on the description, notation, and
      scoring of some perceptual and cognitive
      attributes of the physical environment. In #29
      (Proshansky et al., 1970), pp. 593-618.

   An attempt to devise a systematic notation and analytic
schema for the description of architectural space.

105. Weinstein, N. D. The statistical prediction of
      environmental preferences: Problems of validity
      and application Environment and Behavior, 1976, 8,
      611-626.

   Problems in the application of multiple regression to
the prediction of responses from sets of environmental
variables are noted, relating to the role of chance
associations when large numbers of predictor variables are
combined, as well as problems of group aggregation, and the
dependence of correlation on range.

106. Danford, S. Reliability and validity of selected
      responses to the visual properties of an
      architectural environmental study. Dissertation
      Abstracts International, 1975, 35-B, 5613.

See also: citations 224, 296, 616, 644, 651, 672.

II. PSYCHOLOGICAL PROCESSES IN THE INDIVIDUALS'S
RESPONSE TO THE ENVIRONMENT

3. ENVIRONMENTAL COMPREHENSION AND EVALUATION

This section includes material frequently designated
under the rubric of "environmental perception" (especially
by those outside of the field of psychology), a term used in
reference to very diverse problems, processes and phenomena.
We have differentiated four categories:

a. Environmental Perception. This heading is used in a
narrow sense to refer to works dealing specifically with the
manner in which environmental stimuli, spaces, scenes,
landscapes, etc., are processed through the perceptual
system. Thus the chief focus in this literature is on
judgments of perceived distance, spatiousness, brightness,
and other perceptual qualities of the environment. This is
not a live area of research on the part of environmental
psychologists, but may gain impetus from the recent
publication of Gibson's treatment of perception from an
ecological point of view (item #111).

b. Environmental Cognition. Included under this
heading are works on the popular topics of mental maps,
spatial cognition, orientation in and navigation through
geographic space, etc., which has developed into a
flourishing research area, with major contributions from
such diverse fields as psychology (and especially
developmental psychology), geography and architecture.

c. Environmental Meaning: Subjective Dimensions of
Environmental Stimuli. A considerable amount of research
has been devoted to the determination of the dimensions used
by individuals, either implicitly or explicitly, to
categorize, describe or compare particular environments,
from rooms and houses to towns, neighborhoods, states and
other geographic units. This work has relied on such
diverse techniques as multi-dimensional scaling, factor
analysis, and Repertory-Grid analysis, largely in a
descriptive vein.

d. Environmental Aesthetics: Affective, Evaluative and
Preferential Responses to Environments. This has been an
active area of research, particularly with reference to
stimuli from the natural environment. Even more than in the
realm of environmental cognition, it has served as a point

of convergence for persons from diverse backgrounds,
including psychologists, architects, landscape architects,
individuals from the fields of natural recreation, forestry,
natural resource management, and others. Much of this work
has been descriptive and a-theoretical, but the influence of
psychological theories, notably that of Berlyne, appears to
be increasingly in evidence.

a. Environmental Perception

107. Bartley, S. H. PERCEPTION IN EVERY-DAY LIFE. New
        York: Harper and Row, 1972. 325 pp.

    Application of classical principles of sensation and
perception to phenomena of perception in everyday life, many
of them drawn from the realm of environmental perception.
Includes consideration of all the senses, with particular
emphasis on touch, and on problems of orientation in space.
Affective responses to sensory stimuli are stressed.

108. Birren, F. LIGHT, COLOR AND ENVIRONMENT. New York:
        Van Nostrand & Reinhold, 1969. 131 pp.

    A general treatment of the role of color in our
environment, and of physiological and psychological effects
of color on the individual.

109. Brunswik, E. PERCEPTION AND THE REPRESENTATIVE DESIGN
        OF EXPERIMENTS. Berkeley, California: University
        of California Press, 1956. 154 pp.

    An unorthodox treatment of the psychology of perception
that is of considerable relevance for environmental
psychology through its emphasis on the role of probabilistic
cues in our interpretation of environmental information, as
well as on the need for representative sampling of
environmental stimuli.

110. Gibson, J.J. THE PERCEPTION OF THE VISUAL WORLD.
        Boston: Houghton-Mifflin, 1950. xii, 240 pp.

    An early statement of Gibson's theory of visual
perception, with emphasis on the texture-gradient theory of
space perception.

111. Gibson, J. J. THE ECOLOGICAL APPROACH TO VISUAL
        PERCEPTION. Boston: Houghton-Mifflin, 1979.
        xiv, 332 pp.

    A systematic description of the visually perceived
environment and of the information contained therein,
presenting Gibson's theory of "affordances," and of
information pickup.

112. Hayward, D. G. The psychology and physiology of light
        and color as an issue in the planning and managing
        of environments: A selected bibliography.
        Monticello, Ill.: Council of Planning Librarians,
        1972. (Exchange Bibliographies No. 288). 14 pp.

    An unannotated compilation of general material on
effects of light and color on behavior.

113. Kennedy, J. M. A PSYCHOLOGY OF PICTURE PERCEPTION.
        San Francisco: Jossey-Bass, 1974. xiii, 174 pp.

    An application and extension of the theory of perception
of James Gibson to the perception of pictures, with
particular reference to outline and schematic drawings.

114. Perception and environment. Urban Ecology, 1979, 4,
        95-161.

A special issue which includes papers on sensory,
informational and social aspects of environmental
perception, as well as perceived environmental quality,
perception in traffic, and perception of traffic noise.

115. Ross, H. E. BEHAVIOR AND PERCEPTION IN STRANGE
          ENVIRONMENTS. London: Allen & Unwin, 1974.   171
          pp.

Phenomena of visual and auditory perception in special
environments, such as under-water, as well as in diverse
situations in everday experience, particularly outdoors.

116. Battro, A. M., & Ellis, E. J. L'estimation
          subjective de l'espace urbain. Annee
          Psychologique, 1972, 72, 9-52.

Architecture students were found to underestimate the
size of urban (outdoor) spaces. It was found that training
may result in a compensatory tendency to overestimate.

117. Battro, A. M., Mazzotti, T. B., & Cabral, R. J. G.

          Vers une psycho-physique a l'echelle urbaine.
          Annee Psychologique 1975, 75, 147-152.

A study of people's sense of the size of public squares
in cities; subjects were asked to explore a public square in
their home town, and subsequently to reproduce it by marking
out its dimensions in an open field. Under-estimation was
found to be the rule, particularly as actual size increased.

118. Carr, S., & Schissler, D. Perceptual selection and
          memory in the view from the road. Environment and
          Behavior, 1969, 1, 7-35.

A study of the visual selection of elements visible from an urban expressway by drivers and passengers, both those unfamiliar with the road as well as regular commuters. Memory for different elements was found to be predicted by a joint function of their relative salience and amount of time they were in view.

119.   Garling, T.   Studies in visual perception of
         architectural spaces and rooms: I. Judgment
         scales of open and closed space.   Scandinavian
         Journal of Psychology, 1969, 10, 250-256.   II.
         Judgments of open and closed space by category
         rating and magnitude estimation.   Ibid., 1969, 10,
         257-268.   III.   A relation between judged depth
         and judged size of space.   Ibid., 1970, 11,
         124-131.   IV.   The relation of judged depth to
         judged size of space.   Ibid., 11, 133-145.   V.
         Aesthetic preferences.   Ibid., 1972, 13, 222-227.

A series of programmatic studies on the perception of architectural space in regard to size, depth, openness, etc. The final report extends the research to the realm of aesthetic preferences, and their relation to factors of variation, openness and other attributes.

120.   Hayward, S. C., & Franklin, S. S.   Perceived openness-
         enclosure of architectural space.   Environment and
         Behavior, 1974, 6, 37-52.

The degree of openness of an architectural space was found to be a function of the relation of the height of the space to its depth.

121.   Ittelson, W. H.   Perception of the large-scale
         environment.   Transactions of the New York Academy
         of Sciences, (Series 2), 1970, 32, 807-815.

Discusses several features of environmental perception, including the attribute of the environment as surrounding

the individual, its multi-modal character, the role of
peripheral sensory information, the excess of potential
information contained in the environment in comparison with
the individual's ability to process it, and the role of
action as intrinsically related to perception.

122.   Ittelson, W. H.   Environmental perception and
           contemporary perceptual theory.   In #134
           (Ittelson), pp. 1-19.

   An examination of perception of the physical environment
from the perspective of the transactional view of
perception.

123.   Mackworth, N. H., & Morandi, A. J.   The gaze selects
           informative details within pictures.   Perception
           and Psychophysics, 1967, 2, 547-552.

   A demonstration of the tendency of the human perceiver
to focus preferentially on the most information-laden
portions of pictorial material such as aerial photos.
Limitations to this general principle are included.

124.   Pyron, B. Form and space diversity in human habitats:
           Perceptual responses.   Environment and Behavior,
           1971, 3, 382-411.

   Housing complexes varying in diversity of form and
spatial layout were presented on film.   The extent of
visual exploration increased with the diversity of both
types.   Accuracy of identification of the component elements
(the houses) was mainly a function of the type of spatial
layout, and was highest for compact court-type complexes.

125.   Sadalla, E. K., & Burroughs, J.   Mobile homes in
           hyper-space: Recognition memory for architectural
           form. Environmental Psychology and Nonverbal
           Behavior, 1978, 2, 195-205.

Used multidimensional scaling to determine the relative degree of distinctiveness of each of a set of 30 pictorially represented mobile homes. The resulting measures of distinctiveness were found to be correlated with the ease with which any given mobile home was remembered over a one-week period. (A recognition procedure was used.)

126. Southworth, M. The sonic environment of cities. Environment and Behavior, 1969 1, 49-70.

A study of the role of sounds in an urban environment in people's orientation to that environment and in the affective satisfaction derived from it.

See also: citations 104, 456, 914, 1392, 1399, 1470a.

b. Environmental Cognition

127. Boice, L. P. Encountering a city: The spatial learning process of urban newcomers. Monticello, Ill.: Council of Planning Librarians, 1977. (Exchange Bibliographies No. 1264). 47 pp.

An annotated bibliography covering aspects of migration, environmental adaptation, spatial learning and cognition, and attitudes, with particular reference to newcomers.

128. Canter, D. THE PSYCHOLOGY OF PLACE. New York: St. Martin's, 1977. x, 198 pp.

An integrative account of theory and research on spatial cognition, including descriptive and evaluative aspects of place, cognition of distances and routes, and "cognitive cartography."

129. Downs, R. M., & Stea, D. (Eds.) IMAGE AND
        ENVIRONMENT: COGNITIVE MAPPING AND SPATIAL
        BEHAVIOR. Chicago: Aldine, 1973. xi, 439 pp.

     Contributions from diverse disciplines, including
material on cognitive distance, mapping, orientation,
spatial preference, and developmental aspects.

130   Downs, R. M., & Stea, D. MAPS IN MINDS: REFLECTIONS
        ON COGNITIVE MAPPING. New York: Harper and Row,
        1977. 284 pp.

     A non-technical introduction to mental maps, with many
examples from historical sources, as well as from everyday
experience. Much illustrative material.

131. Golledge, R. G. & Rushton, G. (Eds.) SPATIAL CHOICE
        AND SPATIAL BEHAVIOR: GEOGRAPHIC ESSAYS ON THE
        ANALYSIS OF PREFERENCE AND PERCEPTIONS. Columbus,
        Ohio: Ohio State University Press, 1976. xiii,
        321 pp.

     A collection of papers on mental mapping and analysis of
spatial and geographic preference, emphasizing mathematical
models, measurement problems and theoretical issues.

132. Gould, P. R., & White, R. R. MENTAL MAPS.
        Harmondsworth, England: Penguin, 1974. 204 pp.

     Describes programmatic research on the spatial structure
of geographic preference, studied in diverse countries, and
in groups of students in different regions of the U.S.A.
Traces relationships between information and preference.

133. Hart, R. CHILDREN'S EXPERIENCE OF PLACE: A
        DEVELOPMENTAL STUDY. New York: Irvington, 1979.
        xxv, 518 pp.

A study of the cognition and experience of place and geographic space in school-age children residing in a small rural Vermont village. The research relied on drawings, construction of models, and verbal procedures. Includes reviews of the literature on spatial activity, place knowledge and feelings, and use of geographic space.

134. Ittelson, W. H. (Ed.). ENVIRONMENT AND COGNITION. New York: Seminar Press, 1973. 187 pp.

Papers from a conference, dealing with such topics as environmental perception and cognition, use of projective techniques in geographic research, and environmental design considerations of health-care systems.

135. Liben, L. S., Patterson, A. H., & Newcomb, N. (Eds.) SPATIAL REPRESENTATION AND BEHAVIOR: DEVELOPMENTAL AND ENVIRONMENTAL APPROACHES. New York: Academic Press (in press).

Papers from a conference on spatial representation, including contributions from developmental and environmental psychologists.

135a. Lynch, K. THE IMAGE OF THE CITY. Cambridge, Mass.: MIT Press, 1960. 194 pp.

A classic in the field of cognitive maps, which analyzes the "imageability" of urban environments in terms of structural elements in the lay-out of the city, specifically paths, nodes, landmarks, edges and districts.

136. Moore, G. T., & Golledge, R. G. (Eds.) ENVIRONMENTAL KNOWING: THEORIES, RESEARCH AND METHODS. Stroudsburg, Pa.: Dowden, Hutchinson & Ross, 1976. xxii, 441 pp.

A wide-ranging set of contributions from diverse
disciplines, divided into three parts: I. Theory and
empirical research (including information-processing,
personal constructs, environmental learning and
developmental theory); II. Perspectives (including
sociological, cross-cultural and literary); III.
Methodological.

137. O'Keefe, J., & Nadel, L. THE HIPPOCAMPUS AS A
          COGNITIVE MAP. New York: Clarendon, 1978. xvi,
          570 pp.

A neurological theory of the representation of spatial
information in the brain, arguing for the hippocampus as the
locus for the storage of such information.

138. Piaget, J., & Inhelder, B. THE CHILD'S CONCEPTION OF
          SPACE. New York: Humanities Press, 1956. xii,
          490 pp.

An account, integrating theory and empirical
observation, of development of spatial cognition in the
child, including the gradual recognition of the role of
spatial perspective (i.e., the relativity of the appearance
of an environmental feature to the observer's point of
view).

139. Piaget, J., Inhelder, B., & Szeminska, A. THE CHILD'S
          CONCEPTION OF GEOMETRY. New York: Basic Books,
          1960. 411 pp.

Includes a discussion of the development of children's
comprehension of geographical concepts and relations.

140. Pocock, D., & Hudson, R. IMAGES OF THE URBAN
          ENVIRONMENT. New York: Columbia University
          Press, 1978. x, 181 pp.

A general treatment of mental maps and images of urban environments, including reports of much European research.

141. Acredolo, L. P., Pick, H. L., & Olsen, M. G.
      Environmental differentiation and familiarity as
      determinants of children's memory for spatial
      location. Developmental Psychology, 1975, 11,
      495-501.

Children's ability to recall the location at which a particular event occurred was investigated in school hallways and playgrounds. In an undifferentiated environment, eight year old children were superior to preschool children, but in an environment that was differentiated through the presence of diverse objects the two groups were equivalent.

142. Allen, G. L., Kirasic, K. C., Siegel, A. W., &
      Herman, J. F. Developmental issues in cognitive
      mapping: The selection and utilization of
      environmental landmarks. Child Development, 1979,
      50, 1062-1070.

Compared children's and adults' identification of scenes, that were considered to be high in value as potential landmarks, encountered along a walk, as well as their ability to rank-order the landmarks selected according to distance.

143. Allen, G. L., Siegel, A. W., & Rosinski, R. R. The
      role of perceptual context in structuring spatial
      knowledge. Journal of Experimental Psychology:
      Human Learning and Memory, 1978, 4, 617-630.

Report of a series of experiments on the identification of scenes encountered along a walk, and of the judgment of distances to the scenes. The walk was simulated via slides

presented either in sequential or random order.
Identification and distance estimation were equally accurate
under either method of slide presentation.

144. Andrews, H. F. Home range and urban knowledge of
         school-age children. Environment and Behavior,
         1973, 5, 73-86.

The ability of school children to locate a list of
landmarks from a region of a large metropolitan area
(Toronto) was related to their age, and to the distance of
their school from the target region.

145. Appleyard, D. Styles and methods of structuring a
         city. Environment and Behavior, 1970, 2, 100-117.

Characteristics of mental maps obtained in the new city
of Ciudad Guyana, Venezuela, are related to educational,
occupational as well as other characteristics of the
individual. Distinguishes between sequential and spatial
modes of organization.

146. Banerjee, T., & Lynch, K. On people and places: A
         comparative study of the spatial environment of
         adolescence. Town Planning Review, 1977, 48,
         105-115.

A comparison of adolescents living in one Argentinian
and two Mexican communities, in terms of their environmental
cognition, preference and spatial behavior.

147. Beck, R. J., & Wood, D. Cognitive transformation of
         information from urban geographic fields to mental
         maps. Environment and Behavior, 1976, 8, 199-238.

An analysis of the processes involved in the
individual's representation of an urban environment through

maps, including a discussion of the role of subject-
characteristics, structural characteristics of the
environment, and situational factors.

148. Beck, R., & Wood, D. Comparative developmental
         analysis of individual and aggregated cognitive
         maps of London. In #136 (Moore & Golledge), pp.
         173-184.

   An analysis of a series of maps obtained from a group of
American teenagers visiting London. Maps were drawn at
different times over a six day period following their
arrival, including after training in "Environmental A," a
symbolic mapping language (see pages 351-361 of Moore &
Golledge).

149. Best, G. A. Direction-finding in large buildings. In
         #46 (Canter), pp. 72-75.

   An investigation of people's orientation and way-finding
in buildings, and of the environmental determinants of
"becoming lost."

150. Birnbaum, M. H., & Mellers, B. A. Measurement and the
         mental map. Perception and Psychophysics, 1978,
         23, 403-408.

   A psychophysical model for the subjective metric of
distances among U.S. cities, studied via estimates of ratios
and differences among such distances.

151. Blaut, J. M., McCleary, G. F., & Blaut, A. S.
         Environmental mapping in young children.
         Environment and Behavior, 1970, 2, 335-349.

A study of grade-school children's ability to interpret
information from aerial photographs and to solve simple
problems in "navigating" in a schematically represented
environment.

152. Blaut, J. M., & Stea, D. Mapping at the age of three.
      Journal of Geography, 1974, 73 (7), 5-9.

Studied the precursors of cognitive maps at the nursery
school level, through simple constructional and navigational
exercises in toy-scale models of environments.

153. Briggs, R. Methodologies for the measurement of
      cognitive distance. In #136 (Moore & Golledge),
      pp. 325-334.

Relationships between true and estimated distances
between points in an urban area are found to be curvilinear,
and vary according to direction (toward or away from the
center of the city), as well as shape of the route
(straight-line or angular) connecting the points.

154. Cadwallader, M. T. Cognitive distance in intra-urban
      space. In #136 (Moore & Golledge), pp. 316-324.

A study of the relationship between actual and estimated
distances between points in a metropolitan area. Emphasizes
the linearity of the relationship, but results obtained are
a function of the method of judgment employed (absolute vs.
relative), and are subject to large individual differences.

155. Canter, D., & Tagg, S. K. Distance estimation in
      cities. Environment and Behavior, 1975, 7, 59-80.

Distance estimates between pairs of points within a city
were obtained from residents of cities in different
countries. A general tendency to overestimate was found to

increase with distance in complex cities; the presence of a
prominent river and other salient features both attenuated
this effect and produced other kinds of errors of
estimation.

156. Conklin, H. C. Ethnographic semantic analysis of
          Ifugao landform categories. In #136 (Moore &
          Golledge), pp. 235-246.

     An illustration of the ethnographer's approach to
environmental mapping, combining aerial photography, machine
mapping, field observation and data obtained from
informants, to study the physical and semantic structure of
the landscape in a mountainous area of the Philippines.

157. Cox, K. R., & Zannaras, G. Designative perceptions of
          macro-spaces: Concepts, a methodology, and
          applications. In #130 (Downs and Stea), pp.
          162-181.

     Basic dimensions underlying students' conceptions of the
various states of the United States, as uncovered from
factor analysis of similarity ratings.

158. DeJonge, D. Images of urban areas: Their structure
          and psychological foundations. Journal of the
          American Institute of Planners, 1962, 28, 266-276.

     Lynch's analysis of urban imageability was applied to
the mental maps of residents of different cities in the
Netherlands, and largely confirmed. The phenomena are
related to the laws of organization of the Gestalt school of
psychology.

159. Devlin, A. S. The "small town" cognitive map:
          Adjusting to a new environment. In #136 (Moore &
          Golledge), pp. 58-70.

A group of newcomers to a small city were asked to draw
maps of their new environment two and one-half weeks after
arrival and then three months later. Changes in the
structure and information contained in these maps are
presented and discussed.

160. Downs, R. M., & Stea D. Cognitive maps and spatial
        behavior: Process and products. In #130 (Downs
        and Stea), pp. 8-26.

A brief general introduction to cognitive maps and some
salient issues concerning them.

161. Francescato, D. & Mebane, W. How citizens view two
        great cities: Milan and Rome. In #130 (Downs &
        Stea), pp. 131-147.

A study of the cognitive maps of residents of these two
Italian cities, via the analysis of sketch-maps, utilizing
Lynch's framework. Includes comparisons between native and
non-native, middle and lower class , and male and female
respondents.

162. Fuller G., & Chapman, M. On the role of mental maps
        in migration research. International Migration
        Review, 1974, 8, 491-505.

Gould's methodology for studying the spatial structure
of geographic preference applied to the preference rankings
for the states of the U. S. A. Rankings were obtained from
samples of Asian, Pacific Island and U. S. residents.

163. Golledge, R. G. Methods and methodological issues in
        environmental cognition research. In #136 (Moore
        & Golledge), pp. 300-313.

Consideration of problems of observation, analysis of external representation, use of indirect judgment techniques, and historical reconstructions, as methodological approaches to the study of environmental cognition.

164. Golledge, R. G., Rivizzigno, V. L., & Spector, A. Learning about a city: Analysis by multidimensional scaling. In #131 (Golledge & Rushton), pp. 95-118. Pp. 95-118.

Application of multidimensional scaling to the study of spatial cognition, with particular reference to structuring of urban space.

165. Golledge, R. G., & Spector, A. N. Comprehending the urban environment: Theory and practice. Geographical Analysis, 1978, 10, 403-426.

A developmental model for the representation of urban space is proposed and partially tested through judgments of distances between points in the city of Columbus, Ohio. A structural-developmental account of cognitive learning, elaborated to incorporate the role of social and contextual determinants, is presented.

166. Gould, P. R. The black boxes of Jonkoping: Spatial information and preference. In #130 (Downs & Stea), pp. 235-245.

A study of Swedish children's information about different regions of their country as well as their preferences for them. Discusses the interrelationship between information and preference.

167. Gould, P. R. On mental maps. In #130 (Downs & Stea), pp. 182-220

An overview of research on the spatial structure of
images of different countries and different portions of the
United States, as reflected in preference judgments obtained
from residents of particular countries and states.

168. Gulick, J. Images of an Arab city. Journal of the
        American Institute of Planners, 1963, 29, 179-198.

An attempt to apply Lynch's concepts of urban images to
the cognitive maps of university students from a Lebanese
city. Lynch's exclusively physical criteria for strong
urban images are criticized.

169. Hardwick, D. A., McIntyre, C. W., & Pick, H. L. The
        content and manipulation of cognitive maps in
        children and adults. Monographs of the Society
        for Research in Child Development, 1976, 41 (3,
        Whole No. 166).

An experimental investigation of children's and adults'
sense of direction in an indoor environment (a large room).
Children are found to have accurate cognitive maps, but
first graders have difficulty manipulating them when asked
either to imagine the room rotating or themselves moving
inside the room.

170. Hart, R. A., & Moore, G. T. The development of
        spatial cognition. In #130 (Downs & Stea), pp.
        246-288.

A comprehensive review of theory and research on the
development of the child's concept of space. Stresses the
work of Piaget and Werner.

171. Hazen, N. L., Lockman, J. J., & Pick, H. L.
        Development of children's representations of
        large-scale environments. Child Development,
        1978, 49, 623-636.

A study of six- to nine-year old children's
representations of a complex indoor environment which they
experienced by walking through a series of small
artificially constructed rooms. Measures obtained included
route reversal, landmark-sequence reversal, inference to
relations among elements in the environment, and
reproduction via a model.

172. Heft, H. The role of environmental features in route-
        learning: Two exploratory studies of way-finding.
        Environmental Psychology and Nonverbal Behavior,
        1979, 3, 172-185.

Studied wayfinding in a residential environment and a
biological preserve to determine the respective roles of
landmarks as opposed to geographic orientation as a basis
for route learning in the outdoor environment.

173. Herman, J. F., Kail, R. V., & Siegel, A. W. Cognitive
        maps of a college campus: A new look at freshman
        orientation. Bulletin of the Psychonomic Society,
        1979, 13, 183-186.

An examination of freshmen's knowledge of the spatial
layout of their campus, investigated after three weeks,
three months, and six months on campus. Reports improvement
up to three months in terms of landmark, route and
configuration knowledge. Sex differences in landmark
knowledge are noted.

174. Herman, J. F., & Siegel, A. W. The development of
        cognitive mapping of the large-scale environment.
        Journal of Experimental Child Psychology, 1978,
        26, 389-406.

Traces changes with practice, as well as with age, in
the ability of children between five and ten years to
reproduce the spatial layout of a model town after having

walked through it. Young children's performance was greatly
superior in a bounded as compared to an unbounded space,
while that of older children was equivalent in either type
of space.

175. Holahan, C. J., & Dobrowolny, M. B. Cognitive and
       behavioral correlates of the spatial environment:
       An interactional analysis. Environment and
       Behavior, 1978, 10, 317-333.

   Emphases and distortions in maps drawn by students of
their campus were found to be related to preferential
spatial behavior patterns obtained through verbal self-
report and collective observation.

176. Howard, R. B., Chase, S. D., & Rothman, M. An
       analysis of four measures of cognitive maps. In
       #37 (EDRA 4, Vol. I), pp. 254-264.

   A comparison of distance judgments in geographic space
obtained by the methods of ratio estimation, absolute
judgment, reproduction and modeling. All four methods are
shown to be highly reliable, and satisfactory in terms of
construct validity, yielding closely comparable results.

177. Jones, M. M. Urban path-choosing behavior: A study
       of environmental clues. In #36 (EDRA 3, Vol. I),
       Sec. 11.4, pp. 1-10.

   The role of different types of clues in locating the
entrance to an urban expressway is studied via
cinematographic simulation.

178. Kaplan, R. Way-finding in the natural environment.
       In #136 (Moore & Golledge), pp. 46-57.

Processes of orientation in a park area, studied in junior high school and college students via diverse methods including questionnaires, sketch maps and games.

179. Kaplan, S. Adaptation, structure, and knowledge. In #136 (Moore & Golledge), pp. 32-45.

A brief theoretical analysis of major cognitive mechanisms involved in the process of obtaining knowledge of the environment, including object recognition, information processing, coding of sequential information and abstraction.

180. Kaplan, S. Cognitive maps in perception and thought. In #130 (Downs & Stea), pp. 63-78.

A discussion of some functional aspects of cognitive maps, stressing their evaluative side and their role in adaptive decision-making.

181. Kaplan, S. Cognitive maps, human needs, and the designed environment. In #37 (EDRA 4, Vol. I), pp. 275-283.

Stresses the place of human needs in environmental design, the success of which is related to the satisfaction of the following criteria: legibility, novelty or uncertainty, and opportunity for choice afforded by the environment.

182. Kosslyn, S. M., Pick, H. L., & Fariello, G. R. Cognitive maps in children and men. Child Development, 1974, 45, 707-716.

Proximity rankings were obtained from pre-school children and from adults for a set of objects whose locations in a room had been learned previously. For both

groups the presence of an opaque barrier separating a pair
of objects tended to increase the remembered distance
between them. For the children this effect also applied to
transparent barriers.

183. Ladd, F. C. Black youths view their environment:
        Neighborhood maps. Environment and Behavior,
        1970, 2, 74-99.

    A study of the spatial structure of a Boston
neighborhood as represented through maps by black
adolescents residing in it. Differentiates among pictorial,
schematic, diagrammatic and map-like maps (with and without
landmarks).

184. Lee, T. R. Urban neighborhood as a social-spatial
        schema. Human Relations, 1968, 21, 241-268.

    A detailed analysis of neighborhood maps in relation to
spatial density, siting of shops and amenity services, etc.
Devises the "Neighborhood Quotient," the ratio of the number
of elements (such as houses, shops, etc.) reproduced on a
map to the number actually present, as a measure of
neighborhood differentiation.

185. Lee, T. R. Perceived distance as a function of
        direction in the city. Environment and Behavior,
        1970, 2, 40-51.

    Establishes the existence of the phenomenon of the
foreshortening of estimated distance in the direction
towards the center of the city, rather than away from the
center.

186. Lee, T. R. Psychology and living space. In #130
        (Downs & Stea), pp. 87-114.

A review of the author's research on diverse aspects of spatial cognition and behavior, including concepts of neighborhoods, and the role of spatial distance in social behavior. The place of such data in the work of the architect and designer is discussed.

187. Lord, F. A study of spatial orientation in children. Journal of Educational Reseach, 1941, 34, 481-505.

Children's sense of orientation and direction for both nearby and far away points was studied in school children from grades IV through VIII. The author differentiates between two different frames of reference suggested by his data: one, based on direct experience, for nearby space, the other, based on conventional maps, for further away points.

188. Lowenthal, D. Geography, experience and imagination: Towards a geographical epistemology. Annals of the Association of American Geographers, 1961, 51, 241-260.

An essay on world views, both general and geographical, emphasizing the personal and private character of world views, and examining their anthropocentric character, mutability, variation across cultural groups, and relationship to reality.

189. Lowenthal, D. Research in environmental perception and behavior: Perspectives on current problems. Environment and Behavior, 1972, 4, 333-342.

A brief review of definitional, conceptual and methodological issues in the field of "environmental perception," i.e., cognitive, evaluative and behavioral responses to visually presented environments.

190. Lowrey, R. A. Distance concepts of urban residents. Environment and Behavior, 1970, 2, 52-73.

A methodological study of distance estimates differentiating between conditions that elicit ordinal rather than those that elicit ratio-scale values on scales of subjective distance.

191. Lueck, V. M. Cognitive and affective components of residential preferences for cities: A pilot study. In #131 (Rushton & Golledge), pp. 273-302.

An application of multidimensional scaling to study "cognitive" aspects of verbally designated cities (e.g., dimensions of meaning associated with them) in conjunction with preference responses.

192. Milgram, S., & Jodelet, D. Psychological maps of Paris. In #30 (Proshansky et al., 1976), pp. 104-124.

Parisians' cognitive maps of their city studied through sketch maps supplemented by data on recognition of particular features from photographs, on residential preference for different sections of the city, and on meanings and feelings attached to these sections.

193. Moore, G. T. Theory and research in the development of environmental knowing. In #136 (Moore & Golledge), pp. 138-164.

An overview of findings and thinking on the development of children's representation of their environment from the perspective of cognitive-developmental theory. Particular reference is made to the work of Piaget and Werner and their students and disciples.

194. Moore, G. T. Knowing about environmental knowledge:
     The current state of theory and research on
     environmental cognition. Environment and
     Behavior, 1979, 11, 33-70.

     A review of theory and research, emphasizing the role of
meaning and symbolism in environmental cognition, individual
differences and developmental changes.

195. Moore, G. T., & Golledge, R. G. Environmental
     knowing: Concepts and theories. In #136 (Moore &
     Golledge), pp. 3-24.

     An introduction to the conceptual, historical and
epistemological issues in the field of environmental
cognition.

196. Nagy, J. N. & Baird, J. C. Children as environmental
     planners. In #10 (Altman & Wohlwill, 3), pp.
     259-294.

     Relates a program of research on how children approach
the construction of an ideal village in terms of the spatial
patterning of the elements of the town. The data are
analyzed developmentally, and compared to the children's
representation of their own home town.

197. Nahemow, L. Research in a novel environment.
     Environment and Behavior, 1971, 3, 81-102.

     A descriptive account of exploratory research on
individual's familiarization with, and structuring of, a
novel indoor environment featuring vibrating mirrors,
flashing strobe-lights, etc.

198. Olshavsky, R. W., MacKay, D. B., & Sentell, G.
      Perceptual maps of supermarket locations. Journal
      of Applied Psychology, 1975, 60, 80-86.

      Studied actual and imaged locations of supermarkets in
an urban area via a multidimensional scaling analysis of
proximity ratings obtained for the markets. Both the imaged
and the actual distances were found to be highly correlated
with shopping frequency, the former more so than the latter.

199. Orleans, P. Differential cognition of urban
      residents: Effects of social scale on mapping.
      In #130 (Downs & Stea), pp. 115-130.

      An analysis of differences in cognitive maps of the Los
Angeles area drawn by residents of different sections of the
area, and of differing socio-economic background.

200. Pailhous, J. Elaboration d'images spatiales et de
      regles de displacement. Travail Humain, 1971, 34,
      299-324.

      Studied the progressive structuring of geographic space
that takes place as persons unfamiliar with an environment
are taken on an extended trip through that environment.
Among the findings was a gradual establishment of a
schematic network linking the elements of the environment,
and a dropping out of metric in favor of topological
relations as a function of the formation of this network.

201. Pearce, P. L. Mental souvenirs: A study of tourists
      and their city maps. Australian Journal of
      Psychology, 1977, 29, 203-210.

      Sketch maps of tourists visiting Oxford, England were
scored for number of landmarks, districts and paths and
overall accuracy, and were related to sex, length of stay
and peripheral vs. central accommodation.

202. Rand, G. Some Copernican views of the city.
        Architectural Forum, 1969, 13 (2), 76–81.

      Describes certain features of the cognitive maps of taxi
drivers, stressing the contrast between their ability to
find routes between two points and their difficulty in
representing relationships between points in different
routes.

203. Rapoport, A. Environmental cognition in cross-
        cultural perspective. In #136 (Moore & Golledge),
        pp. 220–234.

      A cognitive-anthropological view of environmental
cognition built on the premise that environment and its
design reflect an individual's or social or cultural group's
mode of organization of time, meaning, communication, and
especially space.

204. Ritchie, J. E. Cognition of place: The island mind.
        Ethos, 1977, 5, 187–194.

      On the basis of anthropological observations of island
residents in Hawaii, Guam and New Zealand, the author argues
that important differences exist between the character of
cognitive maps of islanders as opposed to mainland
residents, the former being more definite, localized and
objective.

205. Saarinen, T. F. Student views of the world. In #130
        (Downs & Stea), pp. 148–161.

      Major features of sketch maps of the world drawn by
students in different countries are compared.

206. Saarinen, T. F. Environmental perception. In I. R.
      Manners & M. W. Mikesell (Eds.), PERSPECTIVES ON
      ENVIRONMENT. Washington, D. C.: Association of
      American Geographers, 1974. Pp. 252-289.
      (Publication No. 13.)

A survey of certain aspects of environmental cognition,
with particular reference to the geographic environment at
the urban and national scale.

207. Sherman, R. C., Croxton, J., & Giovanatto, J.
      Investigating cognitive representations of spatial
      relationships. Environment and Behavior, 1979,
      11, 209-226.

Compared students' spatial representation of their
campus based on distance estimates with that obtained from
construction of maps. The two techniques yielded highly
discrepant results; the former was considered to provide a
more direct and reliable evidence of the features of a
person's cognitive map.

208. Siegel, A. W., Kirasic, K. C., & Kail, R. V., Jr.
      Stalking the elusive cognitive map: The
      development of children's representations of
      geographic space. In #10 (Altman & Wohlwill, 3),
      pp. 223-258.

A review of theoretical concepts and experimental
research, including discussion of neurological bases for
spatial knowledge, development of landmark knowledge, route
learning, and knowledge of spatial configurations.

209. Siegel, A. W., & Schadler, M. The development of
      young children's spatial representations of their
      classrooms. Child Development, 1977, 48, 388-394.

A study of kindergarten children's representation of their classrooms through models, demonstrating changes with experience over the school year and the role of landmarks provided as an aid.

210. Siegel, A. W., & White, S. H. The development of
      spatial representations of large-scale
      environments. In H. W. Reese (Ed.) ADVANCES IN
      CHILD DEVELOPMENT AND BEHAVIOR (Vol. 10). New
      York: Academic Press, 1975. Pp. 9-55.

A review of basic theoretical literature and research on spatial representation and its mechanisms.

211. Stea, D. Program notes on a spatial fugue. In #136
      (Moore & Golledge), pp. 106-120.

A somewhat personalized and refreshingly informal collage of thoughts and ideas on diverse aspects of environmental cognition.

211a. Stevens, A., & Coupe, P. Distortions in judged
      spatial relations. Cognitive Psychology, 1978,
      10, 422-437.

A report of a study on distortions in judgment of spatial relations between two points in geographic space arising when the two points are contained in different geographical or political units.

212. Tolman, E. C. Cognitive maps in rats and men.
      Psychological Review, 1948, 55, 189-208.

A classical paper by the foremost representative of cognitive learning theory, postulating cognitive structures such as maps as mediators of spatial learning.

213. Trowbridge, C. C. On fundamental methods of
         orientation and "imaginary" maps. Science, 1913,
         88, 888-896.

      An early investigation of people's cognitive maps,
studied in terms of people's sense of direction of
particular localities in relation to their home locality
(New York City). Differences among individuals in this
ability appeared to be related to their ability to orient
themselves in the environment generally.

214. Tuan, Y. F. Images and mental maps. Annals of the
         Association of American Geographers, 1975, 65,
         205-213.

      Differentiates between the role of images, schemata, and
mental maps in the individual's representation of the
spatial environment.

215. Wofsey, E., Rierdan, J., & Wapner, S. Planning to
         move: Representing the currently inhabited
         environment. Environment and Behavior, 1979, 11,
         3-32.

      Compared certain characteristics of the cognitive maps
of their urban environment produced by students who were
contemplating an imminent move to a different environment
and those who had no such plans.

216. Zannaras, G. The relation between cognitive structure
         and urban form. In #136 (Moore & Golledge), pp.
         336-350.

      A comparison of the salient cues utilized in navigating
in different urban environments (represented via two- and
three-dimensional maps), as a function of the structure of
the city. Emphasizes the role of structural attributes of
the layout of the city, rather than an individual's urban
experience and other variables relating to the person.

217. Baum D. R. Cognitive maps: The mental representation
      of geographic distance. Dissertation Abstracts
      International, 1978, 38-B, 5606-5607.

218. Gellman, L. H. A chronometric analysis of landmark
      features in the cognitive map. Dissertation
      Abstracts International, 1977, 38-B, 2397-2398.

219. Hansvick, C. L. Comparing urban images: a
      multivariate approach. Dissertation Abstracts
      International, 1978, 38-B, 5093.

See also: citations 605, 608, 619, 885, 888, 1194, 1378,
1379, 1380, 1396, and various items in Ch. 12-d.

c. Environmental Meaning: Subjective Dimensions of
Environmental Stimuli

220. Calvin, J. S., Dearinger, J. A., & Curtin, M. E. An
      attempt at assessing preferences for natural
      landscapes. Environment and Behavior, 1972, 4,
      447-470.

     A factor analysis of semantic differential ratings of 15
natural landscapes yielded two major factors, one identified
as natural scenic beauty, the other as "natural force"
(i.e., potency).

220a. Canter, D. An intergroup comparison of connotative
      dimensions in architecture. Environment and
      Behavior, 1969, 1, 37-48.

Presents the results of two factor analyses of semantic
differential ratings of buildings. The first used
architecture students shown schematic plans of 20 buildings;
the second used students from other disciplines, shown
drawings of 24 interiors. "Coherence" and "friendliness"
emerged as primary factors in both analyses.

221. Downs, R. M. The cognitive structure of an urban
         shopping center. Environment and Behavior, 1970,
         2, 13-39.

A study of the dimensions of connotative meaning
attached to the shops of a shopping center, utilizing factor
analysis of semantic differential ratings.

222. Garling, T. The structural analysis of environmental
         perception and cognition: A multidimensional
         scaling approach. Environment and Behavior, 1976,
         8, 385-415.

Semantic-differential ratings obtained from 24
perspective drawings of urban and suburban scenes were
factor-analyzed, and the results compared to the dimensions
generated by subjecting similarity judgments based on a
subset of 12 of these stimuli to multidimensional scaling.
The former yielded an aesthetic and a social-status factor;
the latter a dimension of style and one of type of area.

223. Hall, R., Purcell, A. T., Thorne, R., & Metcalfe, J.
         Multidimensional scaling analysis of interior,
         designed spaces. Environment and Behavior, 1976
         8, 595-610.

Reports results from two separate multidimensional
scaling analyses of similarity ratings obtained from 11
slides of office building foyers.

224. Harman, E. J., & Betak, J. F. Behavioral geography, multidimensional scaling, and the mind. In #131 (Golledge & Rushton), pp. 3-20.

A critical examination of the uses and limitations of multidimensional scaling for the treatment of data from studies of spatial cognition and preference.

225. Harrison, J., & Sarre, P. Personal construct theory in the measurement of environmental images: Problems and methods. Environment and Behavior, 1971, 3, 351-374.

Presents the theory, rationale and methodology used in the application of repertory grid technique to the study of environmental meaning.

226. Harrison, J., & Sarre, P. Personal construct theory in the measurement of environmental images. Environment and Behavior, 1975, 7, 3-58.

An application of the repertory grid technique in studying the categories and dimensions of meaning by which an English city is conceptualized by its residents.

227. Hendrick, C., Martyniuk, O., Spencer, T. J., & Flynn, J. E. Procedures for investigating the effect of light on impression: Simulation of a real space by slides. Environment and Behavior, 1977, 9, 491-510.

Comparison of semantic-differential ratings of rooms under different conditions of lighting, obtained from real and slide simulated interiors.

228. Horayangkura, V. Semantic dimensional structures: A methodological approach. Environment and Behavior, 1978, 10, 555-584.

Dimensions in the perception of residential environments
were obtained through a sorting technique and semantic-
differential ratings using photographic stimuli. Both
multidimensional scaling and factor analysis revealed three
dimensions: evaluation, organization and urbanization.

229. Hudson, R. Images of the retailing environment: An
       example of the use of the repertory grid
       methodology. Environment and Behavior, 1974, 6,
       470-494.

An application of repertory grid technique in studying
the meanings attached to stores of a shopping center.

230. Lowenthal, D., & Riel, M. The nature of perceived and
       imagined environments. Environment and Behavior,
       1972, 4, 189-207.

Examines the clustering of dimensions of meaning
obtained by intercorrelating ratings of (a) a set of
traversed urban environments, and (b) generic, undefined
environments. Results from semantic-differential ratings of
environmental stimuli are shown to be based in part on
purely semantic connotations of particular concepts,
independent of specific environmental referents.

231. Pedersen, D. M. Dimensions of environmental
       perception. Multivariate Experimental Clinical
       Research, 1978, 3, 209-218.

Semantic-differential ratings of diverse generic
environmental settings (large city, small town, forest,
desert, beach) revealed four main factors: evaluative,
spiritual, activity, and aesthetic appeal.

232. Sonnenfeld, J. Multidimensional measurement of
         environmental personality. In #131 (Golledge &
         Rushton), pp. 51-68.

      Presents a model for the conceptualization of
personality as a determinant of the individual's response to
the geographic environment, and brief applications to the
areas of community perceptions, landscape preferences, and
environmental meaning.

233. Stringer, P. Repertory grids in the study of
         environmental perception. In P. Slater (Ed.), THE
         MEASUREMENT OF INTRAPERSONAL SPACE BY GRID
         TECHNIQUE, Vol. 1: EXPLORATIONS OF INTRAPERSONAL
         SPACE. New York: Wiley, 1976. Pp. 183-208.

      A review of applications of repertory grid technique in
environmental cognition, including the study of urban
images, environmental perception, and evaluation of
alternative planning proposals.

234. Ullrich, J. R., & Ullrich, M. F. A multidimensional
         scaling analysis of perceived similarities of
         rivers in Western Montana. Perceptual and Motor
         Skills, 1976, 43, 575-584.

      Multidimensional scaling analysis of similarity
judgements made of 12 river sections revealed two major
dimensions: size (river-width) and natural/developed.

235. Ward, L. M. Multidimensional scaling of the molar
         physical environment. Multivariate Behavioral
         Research, 1977, 12, 23-42.

      An application of two alternative multidimensional
scaling solutions to similarity judgments of a set of
environmental stimuli and to category sorts of the same
stimuli obtained under different functional orientations.

See also: citations 95, 97, 103, 1398.

d. Environmental Aesthetics: Affective, Evaluative and
Preferential Responses to Environments

236. Berlyne, D. E. AESTHETICS AND PSYCHOBIOLOGY. New
York: Appleton-Century Crofts, 1971. xiv, 336
pp.

A broad overview of theory and experimental findings in
the domain of psychological aesthetics from the perspective
of the author's theory of arousal, exploratory behavior and
affective response to environmental stimuli. Although no
specific treatment is given of problems of environmental
aesthetics, the author's analysis is directly relevant and
readily applicable to that realm.

237. Coomber, N. H., & Biswas, A. K. EVALUATION OF
ENVIRONMENTAL INTANGIBLES. Bronxville, N. Y.:
Geneva Press, 1973. 77 pp.

A review of diverse models to assess environmental
externalities, including willingness-to-pay and other
economic models, as well as aesthetic measure and
attractivity models. Extensive bibliography.

238. Craik, K. H., & Zube, E. H. (Eds.) PERCEIVING
ENVIRONMENTAL QUALITY. New York: Plenum, 1976.
310 pp.

Proceedings of a symposium on perceived environmental
quality and its measurement, with specific reference to
residential and institutional as well as recreational

environments, and to perceptual aspects of environmental quality with respect to air, water and sonic aspects (i.e., noise).

239. Tuan Y. F. TOPOHILIA: A STUDY OF ENVIRONMENTAL PERCEPTION, ATTITUDES AND VALUES. Englewood Cliffs, New Jersey: Prentice Hall, 1974. 260 pp.

A far-ranging treatment of people's feelings and values related to their environment, both natural and man-made, from the perspective of a humanistic geographer.

240. U. S. Environmental Protection Agency, Office of Research and Development. AESTHETICS IN ENVIRONMENTAL PLANNING. Washington, D. C.: U. S. Government Printing Office, 1973. (EPA-600/5-73-009)

An overview of basic and applied literature on general and environmental aesthetics and of its place in environmental law, design and planning.

241. U. S. Forest Service. PROCEEDINGS OF THE CONFERENCE ON METROPOLITAN PHYSICAL ENVIRONMENT, SYRACUSE, 1975. Upper Darby, Pa.: Northeastern Forest Experiment Station, 1977. 447 pp. (USDA Forest Service General Technical Report NE-25.)

Contains a section of brief papers on "human perception, social factors and esthetics of metropolitan environment," as well as a section on noise, and the use of vegetation in noise control.

242. Acking, C. A., & Sorte, G. J. How do we verbalize what we see? Landscape Architecture, 1973, 64, 470-475.

A brief account of a research program on verbal assessments of scenes from the natural and man-made environment. Relationships between pleasantness and complexity, between unity and naturalness, and other related findings are presented.

243.   Basch, D. The uses of aesthetics in planning: A
           critical review. Journal of Aesthetic Education,
           1972, 6, 39-55.

Considers and critically evaluates four approaches taken to determining aesthetic value of environmental features as an element for consideration by planners. They are: (1) Survey (public opinion); (2) Market (economics); (3) Perception (study of people's evaluative responses); and (4) Analytic (analysis of aesthetics into component factors).

244.   Brebner, J., Rump, E. E., & Delin, P. A cross-
           cultural replication of attitudes to the physical
           environment. INTERNATIONAL JOURNAL OF PSYCHOLOGY,
           1976, 11, 111-120.

A replication of the study by Winkel et al. (see item #266) with an Australian sample. Factor analysis of responses to statements on a questionnaire reflecting diverse orientations towards the man-made environment confirm Winkel et al's three factors (preference for order and tasteful design, desire for an exciting environment, and dislike of civic construction), along with three additional factors.

245.   Ertel, S. Exploratory choice and verbal judgment. In
           D. E. Berlyne and K. B. Madsen (Eds.), PLEASURE,
           REWARD, PREFERENCE. New York: Academic Press,
           1973. Pp. 115-132.

A theoretical argument and empirical study concerning the important role of the stimulus attributes of symmetry and balance as determinants of aesthetic preference.

246. Estvan, F. J., & Estvan, E. W.   THE CHILD'S WORLD:
         HIS SOCIAL PERCEPTION. New York: Putnam's, 1959.
         xiii, 302 pp.

   Contains a variety of data on the social perception of
rural and urban children in grades I and VI, including
material on perceptual, attitudinal and preferential
responses to pictures of diverse environmental settings,
such as factory, village, dam, poor house, resort, etc.

247. Howard, R. B., Mlynarski, F. C., & Sauer, C. G., Jr.
         A comparative analysis of affective responses to
         real and represented environments. In #36
         (EDRA-3, Pt. 1), sec. 6-6, pp. 1-8.

   A comparison between verbal ratings given to exteriors
and interiors of buildings when presented on slides and when
made on the scene. A regression-to-the-mean phenomenon was
found for the slide judgments (i.e., judgments were less
extreme). There was little difference between black-and-
white and color slides.

248. Kaplan, R. Some methods and strategies in the
         prediction of preference. In #638 (Zube et al.),
         pp. 118-129.

   A brief discussion of theoretical and methodological
issues in the study of environmental preference, dealing
with problems of sampling, selection of raters and rating
methods, and the dimensionalization of preference responses.

249. Kaplan, S. An informal model for the prediction of
         preference. In #638 (Zube et al.), pp. 92-101.

   Emphasizes the role of legibility and "mystery" (promise
of further information) as determinants of preference for
scenes from the physical environment.

250.   Kaplan, S., Kaplan, R., & Wendt, J. S. Rated
        preference and complexity for natural and urban
        visual material. Perception and Psychophysics,
        1972, 12, 354-356.

    An empirical study of ratings of complexity and
preference for urban and natural scenes, demonstrating that
a strong overall preference for natural as opposed to man-
made stimuli overlays a trend for preference to increase
with complexity within each domain.

251.   Kates, R. The pursuit of beauty in the environment.
        Landscape, 1966-67, 16(2), 21-24.

    Argues that "ugliness," and lack of fit of an
environmental feature in its context, is more readily
identifiable and measurable than is "beauty."

252.   Kreimer, A. Environmental preferences: A critical
        analysis of some research methodologies. Journal
        of Leisure Research, 1977, 9, 88-97.

    A methodological critique of environmental perception
and preference research, with particular attention to
problems arising from the use of photographs and the effects
induced through particular photographic techniques.

253.   Laurie, I. C. Aesthetic factors in visual evaluation.
        In #638 (Zube et al.), pp. 102-117.

    A brief discussion of aesthetics and its role in
landscape evaluation.

254.   Leff, H., Gordon, L., & Ferguson, J. Cognitive set
        and environmental awareness. Environment and
        Behavior, 1974, 6, 396-447.

An investigation of the role of specific induced cognitive sets (e.g., to view a scene in an analytic fashion, or to imagine how the scene might be improved) in altering diverse judgmental and evaluative responses to the environment.

255. Leopold, L. B., & Marchand, M. O. On the quantitative inventory of the riverscape. Water Resources Research, 1968, 4, 709-717.

Develops a system for scaling the scenic value of landscapes based on the determination of a uniqueness ratio, reflecting the relative commonness of rated characteristics of a particular landscape (including physical, biological and "human use and interest" attributes) relative to some reference set. The method is applied to a set of 24 valley areas in the vicinity of Berkeley, California.

256. Lewin, K. L. Kriegslandschaft. Zeitschrift fur angewandte Psychologie, 1917, 12, 440-447.

A fascinating early paper by Kurt Lewin (an influential psychological theorist noted for his field theory of behavior), providing a phenomenal account of the change in the response to landscape occurring in a soldier (the author himself) during the course of his advance to the front lines.

257. Lowenthal, D. Not every prospect pleases: What is our criterion for scenic beauty? Landscape, 1962-63, 12(2), 19-23.

Argues for the cultural, historical and contextual relativity of our standards of beauty in the landscape. "What makes one landscape harmonious, another incongruous, is the entire experience of the viewer."

258. Lozano, E. Visual needs in the urban environment.
      Town Planning Review, 1974, 45, 351-374.

      An analysis of the role of an environment's
conduciveness to orientation, and of its variety, as two
aspects of urban scenes that are considered to be of major
importance in determining the enjoyment afforded by such
environments.

259. Maslow, A. H., & Mintz, N. L. Effects of esthetic
      surroundings: I. Initial short-term effects of
      three esthetic conditions upon perceiving "energy"
      and "well-being" in faces. Journal of Psychology,
      1956, 41, 247-254.

      Pictures of faces were rated more positively on scales
of perceived energy and well-being when presented in an
aesthetically pleasing room as opposed to an unattractive,
messy one.

260. Mintz, N. L. Effects of esthetic surroundings: II.
      Prolonged and repeated experience in a "beautiful"
      and an "ugly" room. Journal of Psychology, 1956,
      41, 459-466.

      Two experimenters carrying out the identical experiment
in a "beautiful" as opposed to an "ugly" room worked at a
more deliberate pace in the former, and showed more positive
scores on scales of phenomenal energy and well-being.

261. Pedersen, D. M. Relationship between environmental
      familiarity and environmental preference.
      Perceptual and Motor Skills, 1978, 47, 739-743.

      An extension of an earlier study (see item #231), to
determine relationships between familiarity and other
experiential variables and four factor-analytically derived
dimensions of meaning attached to four types of

environments. The evaluative factor was the only one to
correlate with familiarity.

261a. Pyron, B. Form and space of human habitats.
      Environment and Behavior, 1972, 4, 87-120.

     Complements the author's experimental study of
perceptual judgments made of simulated housing complexes
(item #124) with evaluative judgments of the degree to which
the environments presented filled needs for diversity,
community, privacy, etc., as well as ratings on a set of
semantic-differential scales. Form information emerged as
more potent than space information.

262. Russell, J. A., & Mehrabian, A. Approach-avoidance
      and affiliation as functions of the emotion-
      eliciting quality of an environment. Environment
      and Behavior, 1978, 10, 355-387.

     A study of the relationship between pleasantness and
arousal value of environmental stimuli (photographic slides
of diverse environmental scenes) and the desire to approach
that environment (based on ratings of several approach-
connoting statements), as well as the desire to affiliate
interpersonally in that setting.

263. Schwarz, H., & Werbik, H. Eine experimentelle
      Untersuchung ueber den Einfluss der syntaktischen
      Information der Anordnung von Baukorpern entlang
      einer Strasse auf Stimmungen des Betrachters.
      Zeitschrift fur Experimentelle und Angewandte
      Psychologie, 1971, 18, 499-511.

     Evaluative responses obtained of film strips of
simulated street scenes were found to be inverted-U shaped
functions of amount of information (diversity of specified
features of the houses on the street).

264.   Sherrod, D. R. et al. Environmental attention,
           affect, and altruism. Journal of Applied Social
           Psychology, 1977, 7, 359-371.

     Studied the effects of prior exposure to pleasant vs.
unpleasant environments (either verbally described, or
viewed through slides) on subsequent helping behavior. The
latter was differentially affected only when attention was
specifically focused on the pleasant or unpleasant aspect of
the environment.

265.   Sonnenfeld, J. Equivalence and distortion of the
           perceptual environment. Environment and Behavior,
           1969, 1, 83-100.

     Compared two contrasting populations residing in Alaska
and Delaware on semantic-differential ratings of a set of
concepts relating to the physical and biotic environment.
Differences between groups, and within groups related to
personality dimensions, were discussed in terms of the place
of focal and contextual aspects of the environment for the
person.

266.   Winkel, G. H., Malek, R., & Thiel, P. The role of
           personality differences in judgments of roadside
           quality. Environment and Behavior, 1969, 1,
           199-223.

     An exploration of individual-differences in judgments of
the aesthetic quality of urban roadsides. These judgments
failed to correlate with measures obtained on a
"personality" questionnaire used in previous work on
response to art; closer relationships were obtained with
measures of orientation towards design, and broader
attitudes towards complexity and ambiguity.

267.   Wohlwill, J. F. Amount of stimulus exploration and
           preference as differential functions of stimulus
           complexity. Perception and Psychophysics, 1968,
           4, 307-312.

Using sets of stimuli scaled for complexity (i.e., diversity), taken from the domain of the physical environment as well as from that of non-representative modern art, relationships between complexity and amount of voluntary exploration of a stimulus on the one hand, and ratings of liking on the other were investigated.

268. Wohlwill, J. F. Environmental aesthetics: The environment as a source of affect. In #8 (Altman & Wohlwill, 1), pp. 37-86.

An integrative review of the literature on evaluative responses to environmental stimuli, with emphasis on the role of specific environmental attributes, and consideration of various methodological problems arising in this area of research.

269. Wohlwill, J. F., & Kohn, I. Dimensionalizing the environmental manifold. In #54 (Wapner et al.), pp. 19-45.

A plea for the use of objectively defined dimensions of the physical environment to study functional relationships between specified environmental properties and perceptual or evaluative responses, illustrated by reference to the role of uncertainty in environmental preference, the optimization principle, and the study of adaptation level effects in environmental evaluation.

270. Gifford, R. D. Personal and situational factors in judgments of typical architecture. Dissertation Abstracts International, 1977, 37-B, 6303.

271. Locasso, R. M. The influence of a beautiful vs. an ugly interior on selected behavioral measures. Dissertation Abstracts International, 1977, 37-B, 5859-5860.

Failed to replicate the results of Maslow and Mintz
(#259). The aesthetic appearance of a room generally failed
to influence behavior on a variety of judgment and
performance tasks, except on one involving ratings of
pictures of rooms, in which the judgments were found to be
displaced away from the pole represented by the appearance
of the room.

See also: citations 105, 108, 114, 189, 197, 652, 674, 675,
676, 740, 905, 923, 1222, 1322, 1386, 1388, 1391.

4. ENVIRONMENTAL STIMULATION AND STRESS

This chapter includes material on the general effects of environmental stimulation on behavior and its development, the effects of isolation and special or unusual environments, and the role of specific environmental variables. Among the variables considered are environmental pollutants, heat and cold, noise, etc. Since most of these conditions represent sources of stress for the individual, a section specifically devoted to environmental stress is also included. The chapter is organized in five parts.

a. <u>Environmental Stimulation</u>. Included here is research and theory dealing with the role of environmental stimulation, defined in a generic sense, upon behavioral functioning, with particular emphasis upon child development.

b. <u>Environmental Stress</u>. This section contains general treatments of environmental stress, with particular reference to the urban environment, as well as material which includes, or is relevant for, discussion of the environment as a source of stress. The work of Hans Selye and his postulate of the General Adaptation Syndrome is central to much of this work, as is the more psychologically and cognitively oriented work of Glass and Singer.

c. <u>Specific Environmental Dimensions and Sources of Stress</u>. Particular dimensions of the physical environment such as gravity, temperature, and ionized air may function as stressors under exceptional conditions (e.g. temperature extremes or weightlessness). Environmental pollutants are also covered in this section.

d. <u>Noise</u>. While noise as a specific source of environmental stress is equivalent to the stresses cited

above, it is accorded a separate section, in response to the
very large volume of material in this area. Consequently,
the coverage of this literature is highly selective, with
emphasis placed upon both field research and laboratory work
which is relatively broad in terms of handling of stimulus
and/or behavioral dimensions. A comprehensive annotated
bibliography on the effects of noise has very recently been
published (see item #322). It includes an extensive chapter
on behavioral issues, and thus provides a more complete
reference to work on this topic.

     e. Isolation and Extreme or Unusual Environments. This
section covers material on the effects of isolation, as
studied under both laboratory and field conditions. It
includes work on individuals living in relatively isolated
groups or communities, as well as material on rather special
kinds of environments (e.g. under-water, space, etc.).

a. Environmental Stimulation

272.   Walsh, R. N. & Greenough, W. T. (Eds.), ENVIRONMENTS
       AS THERAPY FOR BRAIN DYSFUNCTION. New York:
       Plenum, 1976. 376 pp.

     A collection of workshop-papers on the role of sensory
stimulation in the etiology of and recovery from brain
damage.

273.   Freeman, H. Mental health and the environment.
       British Journal of Psychiatry, 1978, 132, 113-124.

     A review of literature on relationships between
environmental conditions and mental health, focusing on the
role of density, phenomena of territoriality and personal
space, and problems of urban environments.

274. Heft, H. Background and focal environmental
         conditions of the home and attention in young
         children. Journal of Applied Social Pyschology,
         1979, 9, 47-69.

     The quality, type and amount of environmental stimuli,
both background and focal, afforded to the child in the
home, was assessed via observations and interviews, and
correlated with indices of kindergarten-age children's
performance on selective attention tasks. Consistent
relationships were found for the variable of ambient noise
level in the home.

275. Lipowski, Z. J. The conflict of Buridan's Ass, or
         some dilemmas of affluence: The theory of
         attractive stimulus overload. American Journal of
         Psychiatry, 1970, 127, 273-279.

     Argues that our society is characterized by a surfeit of
attractive stimulation, resulting in conflicts for the
individual having to choose among equally attractive
alternatives.

276. Lipowski, Z. J. Surfeit of attractive information
         input: A hallmark of our environment. Behavioral
         Science, 1971, 16 467-471.

     A paper similar to the preceding.

277. Parke, R. D. Children's home environments: Social
         and cognitive effects. In #10 (Altman & Wohlwill,
         3), pp. 33-82.

     Contains an extensive review of literature on such
aspects of the home environment as toys, TV, noise, privacy
and crowding, and their relationship to the cognitive and
social development of the child.

278.   Russell, R. W.   Behavioral adjustment and the physical
       environment.   International Journal of Psychology,
       1977, 12, 79-92.

       A discussion of psychobiological processes of adaptation
and adjustment to environmental conditions, stressing the
need for consideration of behavioral and biological limits
of tolerance in the design and management of the
environment.

279.   Wachs, T. D.   The optimal stimulation hypothesis and
       early development: Anybody got a match?   In I.
       Uzgiris & J. McV. Hunt (Eds.), THE STRUCTURING OF
       EXPERIENCE.   New York: Plenum, 1978.   Pp.
       153-177.

       Reviews the evidence on the optimal-level-of-stimulation
hypothesis at both the animal and the human levels, in terms
of both preference responses and enhancement of the
individual's development.   In both respects the hypothesis
receives only qualified support from the research to date.
The author stresses the need for greater attention to
individual differences, so that levels of stimulation can be
considered relative to the state and level of development of
the organism at a particular time.

280.   Wachs, T. D.   Proximal experience and early cognitive
       intellectual development: The physical
       environment.   Merrill-Palmer Quarterly, 1979, 15,
       3-41.

       A longitudinal study of the physical-environmental
correlates of the early intellectual development of children
from the age of 11 months to 24 months.   Results point to
the importance of several features of the environment,
notably freedom for exploration, adequacy of personal space,
provision for temporary escape from impinging stimuli, and
control over excess noise.   Effects appeared to be specific
to particular cognitive variables.

281. Wachs, T. D., Francis, J., & McQuiston, S.
      Psychological dimensions of the infant's physical
      environment. Infant Behavior and Development,
      1979, 2, 155-161.

     The home-environments of 136 infants were assessed by
the application of a Home Stimulation Inventory, and the
results factor-analyzed, revealing a set of 13 factors,
among which 10 referred to aspects of the physical
environment and the quality of the stimulation it contains.
Factors accounting for the greatest amount of variance
included variety of stimulation, amount of "traffic" (i.e.,
activity) in the home, and availability of books and toys.

282. Wohlwill, J. F. Human adaptation to levels of
      environmental stimulation. Human Ecology, 1974,
      2, 127-147.

     An overview of research and theory relating to processes
of behavioral adaptation to stimulation in general as well
as the effects of environmental stimulation upon behavior
along such dimensions as intensity, variety and complexity
of stimulation and the operation of adaptation levels with
respect to such dimensions.

283. Wohlwill, J. F. Behavioral response and adaptation to
      environmental stimulation. In #287 (Damon), pp.
      295-334.

     An alternate, slightly expanded version of the
preceding.

284. Wohlwill, J. F., & Heft, H. Environments fit for the
      developing child. In H. McGurk (Ed.), ECOLOGICAL
      FACTORS IN HUMAN DEVELOPMENT. Amsterdam: North
      Holland Press, 1977. Pp. 125-138.

Discusses the role of environmental stimulation in the
development of the child, emphasizing the possible
deleterious effects of high levels of stimulation, and of
background stimulation in particular (e.g., noise).
Includes an account of a study on the relationship between
amount and type of background stimulation in the home
environment and young children's performance on selective-
attention tasks.

285. Wohlwill, J. F., & Kohn, I. The environment as
        experienced by the migrant: An adaptation-level
        view. Representative Research In Social
        Psychology, 1973, 4, 135-164.

An account of a study of individuals who had moved to an
intermediate-sized urban area from either a large
metropolitan area or a rural or small town environment,
focusing in their assessment of attributes of their new
environment.

See also: citations 269, 558, 932, 959, 1043, 1136, 1395,
1473.

b. Environmental Stress

286. Appley, M., & Trumbull, R. (Eds.). PSYCHOLOGICAL
        STRESS: ISSUES IN RESEARCH. New York: Appleton-
        Century Crofts, 1967. 471 pp.

Based on an interdisciplinary conference on
psychological stress. Several chapters deal with
environmental stress, from isolation, captivity, conditions
of danger, etc.

287. Damon, A. (Ed.). PHYSIOLOGICAL ANTHROPOLOGY. New
York: Oxford University Press, 1975. xiii, 367
pp.

A collection of reviews on physiological and behavioral
response and adaptation to environmental conditions,
including high altitude, heat, cold, noise, nutrition, etc.

288. Glass, D., & Singer, J. URBAN STRESS. New York:
Academic Press, 1972. 182 pp.

This volume presents the results of a program of
laboratory research on people's response to environmental
stressors; it does not deal with field studies, as the title
might suggest. Particular attention is given to noise as a
stressor, and the role of predictability of and control over
a stressor in mitigating its effects.

289. Gold, J. R. Stress and the quality of urban life: A
selected bibliography. Monticello, Ill.: Vance
Bibliographies. (Public Administration Series,
Pp-340) 25 pp.

An unannotated bibliography, preceded by a brief
introduction, covering the concept of stress, urban stress
and pathology, physical stressors (mainly noise), crowding
and residential density in relation to pathology, stress and
highrise living, and general material relating to urban
stress.

290. Klausner, S. Z. (Ed.) WHY MAN TAKES CHANCES. New
York: Anchor, 1968. xii, 267 pp.

A small collection of essays on diverse manifestations
of stress-seeking behavior.

291. Lantis, M. Environmental stresses on human behavior:
     Summary and suggestions. Archives of Environmental
     Health, 1968, 17, 578-585.

     An analysis of stress of living in Arctic Eskimo
communities, emphasizing problems of social isolation and
confrontation with a foreign (i.e., Eskimo) culture.

292. Levi, L., & Anderson, L. PSYCHOSOCIAL STRESS:
     POPULATION, ENVIRONMENT, AND QUALITY OF LIFE. New
     York: Spectrum, 1975, 142 pp.

     A general review of demographic and epidemiological data
relating to environmentally based stresses, with particular
reference to population density, migration, and
urbanization.

293. Selye, H. STRESS IN HEALTH AND DISEASE. Boston:
     Butterworths, 1976. 1500 pp.

     An encyclopedic compendium covering the physiological,
psychological and medical literature on stress. Includes a
170-page section on diseases of adaptation and a 230-page
section on the theory of stress. A further section covers
specific stressors, including environmental stressors such
as temperature, sound, climate and occupational sources of
stress such as driving, air traffic control, etc.

294. Baddeley, A. D. Selective attention and performance
     in dangerous environments. British Journal of
     Psychology, 1972, 63, 537-546.

     A review of findings on people's behavioral response to
conditions of danger, in combat and combat-like situations,
as well as in such groups as sport parachute jumpers and
deep-sea divers. The interpretation focuses on effects of

narrowing of attention resulting from enhanced arousal under danger.

295. Carson, D. H. Environmental stress and the urban dweller. Michigan Mental Health Research Bulletin, 1968, 2(4), 5-11.

A brief review of physical and psychological effects of environmental stressors, including pesticides, noise, air pollutants, and overcrowding.

296. Frankenhaeuser, M. The experimental psychology research unit. Man-Environment Systems, 1975, 5, 193-195.

A review of the work being conducted at the Swedish Medical Research Council on physiological and behavioral effects of environmental stress, with particular emphasis on effects of stimulus overload and deprivation, under both laboratory and field conditions.

297. Glass, D. C., Singer, J. E., & Pennebaker, J. W. Behavioral and physiological effects of uncontrollable environmental events. In #13 (Stokols), pp. 131-152.

An account of the authors' programmatic research on response and adaptation to environmental stressors, and of the role of predictability and controllability as mediators of aftereffects from such stressors, extended to consideration of the differential susceptibility of two types of individuals to stress-mediated heart disease.

298. Hulbert, S. F. Drivers' GSR in traffic. Perceptual and Motor Skills, 1957, 7, 305-315.

Galvanic-Skin responses were recorded for three drivers, and correlated with conditions of traffic.

299. Kanner, A., Kafry, D., & Pines, A. Conspicuous in its absence: The lack of positive conditions as a source of stress. Journal of Human Stress, 1978, 4(4), 33-39.

For a group of 84 undergraduate students and 205 employed adults, conditions of work and life were separated into positive and negative (the two being uncorrelated). Absence of positive conditions was highly correlated with dissatisfaction in both work and life, while presence of negative conditions was only moderatedly correlated with dissatisfaction in life and uncorrelated with dissatisfaction in work. A similar pattern was shown for a scale of experienced tedium.

300. Michaels, R. M. Tension responses of drivers generated on urban streets. NRC Highway Research Board Bulletin #271, 1960, 29-64.

Compared galvanic-skin reflex measures of drivers driving over two alternative routes in an urban area.

301. Miller, J. G. Information input overload and psychopathology. American Journal of Psychiatry, 1960, 116, 695-704.

Demonstrated qualitative and quantitative changes in performance in a complex information-processing task as the rate of information-input was increased to produce varying degrees of overload.

302. Ostfeld, A. M., & D'Atri, D. A. Psychophysiological responses to the urban environment. International Journal of Psychiatry in Medicine, 1975, 6, 15-28.

A review of diverse data on psychophysiological effects of crowding and other environmental stressors considered to be relevant to the conditions of stress in urban areas.

303. Rule, B. G. & Nesdale, A. R. Environmental stressors, emotional arousal, and aggression. In I. G. Sarason & & C. D. Spielberger (Eds.), Stress and Anxiety, Vol. 3. Washington, D. C.: Hemisphere, 1976. Pp. 87-103.

A review of relations between environmental variables, such as temperature, noise and density, and aggression, mediated by increase in arousal.

304. Wilkinson, R. Some factors influencing the effect of environmental stressors on performance. Psychological Bulletin, 1969, 72, 260-272.

A review of effects of stress on performance, with particular reference to work environments.

305. Roberts, C. J. An interview study of the experience of urban stress by newcomers and native city dwellers. Dissertation Abstracts International, 1979, 39-B, 3593.

See also: citations 685, 715, 761, 920, 1261, 1265, 1408.

c. Specific Environmental Dimensions and Sources of Stress

306. Burns, N., Chambers, R. M., & Hendler, E. (Eds.), UNUSUAL ENVIRONMENTS AND HUMAN BEHAVIOR. New York: Macmillan, 1963. 438 pp.

An overview of research on environmental stresses
involved in space flight, including weightlessness, high-
altitude, vibration and radiation.

307. National Clearing House for Mental Health Information,
       U.S. National Institute of Mental Health.
       POLLUTION: ITS IMPACT ON MENTAL HEALTH.
       Washington: U. S. Gov. Printing Office, 1972.
       (DHEW Publication No. HSM 72-9135).

A brief review of the effects of pollutants, noise,
housing conditions, recreation and stimulation on mental
health, followed by abstracts of the individual papers
included in the review.

308. William, J. S., Jr., Leyman, E., Karp, S. A., and
       Wilson, P. T. ENVIRONMENTAL POLLUTION AND MENTAL
       HEALTH. Washington, D. C.: Information Resources
       Press, 1973. 136 pp.

Alternate edition of the previous item.

309. Baron, R. A., & Bell, P. A. Aggression and heat: The
       influence of ambient temperature, negative affect,
       and a cooling drink on physical aggression.
       Journal of Personality and Social Psychology,
       1976, 33, 245-255.

Showed that at high temperatures subjects are more prone
to exhibit physical aggression towards others who have given
them positive evaluations.

310. Beard, R. R., & Wertheim, G. A. Behavioral impairment
       associated with small doses of carbon monoxide.
       American Journal of Public Health, 1967, 57,
       2012-2022.

Impairment in the ability of rats and college-aged students to perform temporal discrimination tasks was found to be in direct proportion to carbon monoxide concentration.

311. Bechtel, R. B., & Ledbetter, C. B. The temporary environment: Cold regions habitability (U.S. Army Corps of Engineers Technical Report 76-10). Hanover, New Hampshire: U. S. Army Cold Regions Research and Engineering Laboratory, 1976.

Not available for annotation.

312. Bell, P. A., & Baron, R. A. Aggression and heat: The mediating role of negative affect. Journal of Applied Social Psychology, 1976, 6, 18-30.

A companion study to that of Baron and Bell (item #309).

313. Breisacher, P. Neuropsychological effects of air pollution. American Behavioral Scientist, 1971, 14, 837-864.

A general review of the effects of air pollution on physiological, perceptual and behavioral functioning.

314. Caston, J., Cazin, L., Gribenski, A., & Lannou, J. Pesanteur et comportement (Approche psychophysiologique). Annee Psychologique, 1976, 76, 145-175.

An overview of the effects of gravity on both animals and humans, with emphasis on sensory mechanisms as well as biological functioning and perception of verticality.

315. Cohen, M. M. Sensory motor adaptation and after-
         effects of exposure to increased gravitational
         forces. Aerospace Medicine, 1970, 41, 318-322.

      Exposure to a 2.0 G (twice normal) environment results
in errors on a hand-eye coordination task but the effect is
compensated. Transient after-effects were observed.

316. Hawkins, L. H., & Barker, T. Air ions and human
         performance. Ergonomics, 1978, 21, 273-278.

      A laboratory study of effects of positive and negative
ionization on performance on psycho-motor tasks and cicadian
rhythm. The latter was affected equally by both positive
and negative ionization; performance itself was improved by
negative ionization, but unaffected by positive ionization.

317. Jones, M. H. Pain thresholds for smog components. In
         #14 (Wohlwill and Carson), pp. 61-65.

      A laboratory study, using a combination of
psychophysical and multiple-correlation methodologies, to
determine the threshold for eye irritation corresponding to
particular concentrations of different components of
polluted air.

318. Ramsey, H. R. Human factors and artificial gravity.
         Human Factors, 1971, 13, 533-542.

      Reviews research on effects of artificial gravity
conditions on orientation and performance; implications for
problems of man-in-space are discussed.

319. Ryback, R. S., Trimble, R. W., Lewis, O. F., &
         Jennings, C. L. Psychological effects of
         prolonged weightlessness (bed-rest) in young
         healthy volunteers. Aerospace Medicine, 1971, 42,
         408-415.

Effects of weightlessness were simulated by confining eight airmen to a bed-rest condition for a five-week period. Those not permitted exercise were more anxious, hostile and depressed than those who were.

320. Charry, J. M. Meteorology and behavior: The effects of positive air ions on human performance, physiology and mood. Dissertation Abstracts International, 1977, 37-B, 4751.

See also: citations 71, 604.

d. Noise

321. Bragdon, C. R. NOISE POLLUTION: THE UNQUIET CRISIS. Philadelphia: University of Pennsylvania Press, 1971. xxii, 280 pp.

A general introduction to noise as a psychological hazard and social and community problem. Includes a detailed account of a community noise survey undertaken in Philadelphia.

322. Bragdon, C. R. NOISE POLLUTION: A GUIDE TO INFORMATION SOURCES. Detroit: Gale Research Co., 1979. 524 pp.

A comprehensive annotated bibliography of research on noise, including an extensive chapter on behavioral effects.

323. Broadbent, D. E. PERCEPTION AND COMMUNICATION. New
          York: Pergamon Press, 1958. 338 pp.

     A general theoretical treatment of auditory perception
and communication, emphasizing the selective nature of
attention and the role of filtering in the perception of
speech. A chapter is devoted to the effects of noise on
behavior.

324. Broadbent, D. E. DECISION AND STRESS. New York:
          Academic Press, 1971. xiv, 522 pp.

     An extension of the author's theory of attention and
selective perception (item #323), dealing with the process
of vigilance, effects of noise, and the problem of arousal,
among other topics.

325. Kryter, K. D. THE EFFECTS OF NOISE ON MAN. New York:
          Academic Press, 1970. 639 pp.

     A thorough review of noise research with major emphasis
on laboratory studies of physiological and behavioral
effects.

326. U.S. Environmental Protection Agency. PROCEEDINGS OF
          THE INTERNATIONAL CONGRESS ON NOISE AS A PUBLIC
          HEALTH PROBLEM. Dubrovnik, Yugoslavia, 1973.
          Washington, D.C.:_ U.S. Government Printing
          Office, n.d. (Doc. 550/9-73-008). 815 pp.

     Includes papers on hearing loss, effects on performance
and behavior, physiological and psychological stress
responses, effects on sleep, and community response to noise
(especially from aircraft).

327. Ward, W. D., & Fricke, J. E. NOISE AS A PUBLIC HEALTH
          HAZARD (Proceedings of the Conference, Washington,

D. C., June 13-14, 1968). Washington, D. C.:
American Speech and Hearing Association, 1969.
(ASHA Reports #4). xi, 383 pp.

Included are the effects of noise on hearing thresholds,
physiological and psychological states, and speech
intelligibility. Discusses industrial noise, community
noise, and community control over noise, and special
problems resulting from technology (e.g., sonic booms).

328. Boles, W. R., & Hawyard, S. C. Effects of urban noise
     and sidewalk density upon pedestrian cooperation
     and tempo. Journal of Social Psychology, 1978,
     104, 29-35.

Combined ambient noise and density of pedestrian traffic
were found to be related to increased walking speed for
women and to decreased readiness to cooperate for both
sexes.

329. Borsky, P. N. Effects of noise on community behavior.
     In #327 (Ward & Fricke), pp. 187-192.

A four-step program of research on the effects of noise
in a community is outlined, involving (a) the definition and
measurement of exposure to sound of people in the community,
and the determination of (b) the extent of interference of
the noise with living conditions and everyday activities,
(c) different people's annoyance and irritation due to the
noise, (d) strength of disposition to complain about the
noise, and actual incidence of complaints received. Factors
affecting the last two of these are listed.

330. Broadbent, D. E. Effects of noise on behavior.   In
     C. M. Harris (Ed.), HANDBOOK OF NOISE CONTROL.
     New York: McGraw-Hill, 1957. Pp.10-1 - 10-34.

A review of research on physiological measures of the effects of noise, stimulus correlates of reports of annoyance from noise, and effects of noise on efficiency of human performance.

331.   Bronzaft, A. L., & McCarthy, D. P.   The effect of
          elevated train noise on reading ability.
          Environment and Behavior, 1975, 7, 517-527.

Reading scores of children in classrooms directly adjoining railroad tracks were found to be significantly lower than those for children in classrooms on the opposite side of the same building. Effects of the noisy environment on the children and on the teacher were discussed as likely contributors to this difference.

332.   Cameron, P., Robertson, D., & Zaks, J.   Sound
          pollution, noise pollution, and health: Community
          parameters. Journal of Applied Psychology, 1972,
          56, 67-74.

Families in Detroit and Los Angeles were interviewed concerning the most frequently experienced sources for noise and sound at work and at home.

333.   Cermak, G. W., & Cornillion, P. C.   Multidimensional
          analyses of judgments about traffic noise.
          Journal of the Accoustic Society of America, 1976,
          59, 1412-1420.

The perception of sounds of traffic was studied by obtaining similarity-dissimilarity ratings and preferential choices for pairs of 13 such sounds. Multidimensional scaling of both sets of data showed sound intensity to be the most important determiner of the responses; information about the source of sounds also contributed to the dissimilarity ratings.

334. Cohen, A. Effects of noise on psychological state.
In #327 (Ward & Fricke), pp. 74-88.

A brief review of research on effects of noise on
sensory, perceptual and cognitive processes, and on
affective and motivational states.

335. Cohen, S., Evans, G. W., & Krantz, D. S.
Physiological, motivational and cognitive effects
of aircraft noise on children: Moving from the
laboratory to the field. American Psychologist,
1980, 35, 231-243.

Reports preliminary results from a field study of
children residing and attending schools in the vicinity of
the Los Angeles International Airport. Evidence of both
physiological and motivational effects of noise tended to
corroborate previous findings from laboratory research;
there was, however, little evidence of adaptation to noise
as a function of length of residence, on the part of either
the children or their parents, except at the physiological
(blood-pressure) level.

336. Cohen, S., Glass, D. C., & Phillips, S. Environment
and health. In H. E. Freeman, S. Levine, & L. G.
Reeder (Eds.), HANDBOOK OF MEDICAL SOCIOLOGY.
Englewood Cliffs, N. J.: Prentice Hall, 1979, pp.
134-149.

A brief review of environmental factors related to
physical and psychological health, with primary reference to
effects of noise.

337. Cohen, S., Glass, D. C., & Singer, J. E. Apartment
noise, auditory discrimination, and reading
ability in children. Journal of Experimental
Social Psychology, 1973, 9, 407-422.

A study of children residing in an apartment house built
on top of a major urban highway. Vertical distance of a
child's apartment from the highway was correlated with
auditory discrimination ability, and with reading
achievement in school, but the latter correlation vanished
when the effect of socioeconomic level was partialled out.

338.   Donnerstein, E., & Wilson, D. W. Effects of noise and
       perceived control on ongoing and subsequent
       aggressive behavior. Journal of Personality and
       Social Psychology, 1976, 34, 774-781.

A laboratory study demonstrating the effectiveness of
noise, and the lack of a sense of perceived control of the
noise, in increasing aggressive behavior in emotionally
aroused patients.

339.   Fiedler, F. E., & Fiedler, J. Port noise complaints:
       Verbal and behavioral reactions to airport-related
       noise. Journal of Applied Psychology, 1975, 60,
       498-506.

A survey of residents living in proximity to the
Seattle-Tacoma airport shows some relationship between
objective noise levels and reports of annoyance, but non-
obtrusive measures of behavior and life-style showed no
consistent relationship to noise, nor did length of
residence. Individual factors related to noise tolerance
appeared to be at least as important as objective noise
levels in determining response to environmental noise.

340.   Gardner, G. T. Effects of human subjects regulations
       on data obtained in environmental stressor
       research. Journal of Personality and Social
       Psychology, 1978, 36, 628-634.

In accordance with the findings of Glass and Singer
(#288) and others on the role of perceived control over
noise in mitigating negative after-effects from exposure to

noise, this study found that subjects who had been asked for
their informed consent prior to participating in the study
were less susceptible to such after-effects than those who
had not.

341. Graeven, D. B. The effects of airplane noise on
        health: An examination of three hypotheses.
        Journal of Health and Social Behavior, 1974, 15,
        336-343.

       In a study of over 500 individuals exposed to varying
levels of noise from airplanes, consistent relationships
were found between reported health problems and awareness
of, as well as experienced annoyance from noise. Neither
objective levels of exposure nor knowledge concerning
effects of noise on health showed significant correlations
with incidence of health problems.

342. Graeven, D. B. Necessity, control, and predictability
        of noise as determinants of noise annoyance.
        Journal of Social Pyschology, 1975, 95, 85-90.

       A field questionnaire study of 200 residents of a city
in a large metropolitan area, to determine relationships
between perceived necessity for, predictability of and
control over sources of noise in specific environmental loci
(home, neighborhood, job) and experienced annoyance due to
noise. Low positive relationships were found for necessity
and controllability.

343. Hartley, L. R. Performance during continuous and
        intermittent noise and wearing ear protection.
        Journal of Experimental Psychology, 1974, 102,
        512-516.

       Different measures of task performance (e.g., time gaps,
vs. errors) are differently affected by continuous as
opposed to intermittent noise. Certain effects of the
former may be attributable to monotony, while intermittent
noise may create temporary arousal.

344. Hitchcock, J. & Waterhouse, A. Expressway noise and
       apartment tenant response. Environment and
       Behavior, 1979, 11, 251-267.

       A survey of residents in the vicinity of an expressway
in a metropolitan area. Level of experienced annoyance from
noise and intention to move were found unrelated to socio-
economic factors and relatively impervious to economic
considerations such as amount of rent, or to compensating
convenience of access to the highway.

345. Hockey, G. R. Effect of loud noise on attentional
       selectivity. Quarterly Journal of Experimental
       Psychology, 1970, 22, 28-36.

       Noise was shown to facilitate performance on a primary
task of tracking a stimulus, while interfering with the
detection of a peripheral stimulus constituting a secondary
task. Noise is considered to play a dual role, increasing
arousal while narrowing the span of attention.

346. Hovey, H. B. Effects of general distraction on the
       higher thought processes. American Journal of
       Psychology, 1928, 40, 585-591.

       Studied the effects of a concentrated dosage of diverse
visual and auditory distractors on mental-test performance.
Observed only minimal impairment of test performance, but
reports indirect evidence on the stressful nature of the
experience.

347. Humphrey, C. R., Bradshaw, D. A., & Krout, J. A. The
       process of adaptation among suburban highway
       neighbors. Sociology and Social Research, 1978,
       62, 246-266.

An investigation of modes of coping with discomfort and stress from noise and air pollution experienced by residents living in close proximity to a limited-access highway serving the suburbs of a major metropolitan area.

348. Levy-Lebeyer, C., Vedrenne, B., & Veyssiere, M.
      Psychologie differentielle des genes dues au
      bruit. Annee Psychologique, 1976, 76, 245-256.

A brief review of individual differences in susceptibility to noise, related to age, sex and personality. The authors postulate that negative responses to noise may be in part related to the individual's overall level of environmental dissatisfaction.

349. Lundberg, U., & Frankenhaeuser, M.
      Psychophysiological reactions to noise as modified
      by personal control over noise intensity.
      Biological Psychology, 1978, 6, 51-59.

Subjects given a choice of which of five levels of noise (between 70 and 105 dbA) they would tolerate while working on arithmetic problems displayed lower states of arousal, in terms of both physiological and self-report measures, than those who had no control over the noise. Individual differences in both groups were related to inner vs. outer locus of control.

350. Mathews, K. E., Jr., & Canen, I. K. Environmental
      noise level as a determinant of helping behavior.
      Journal of Personality and Social Psychology,
      1975, 32, 571-577.

Results from two studies, one in the laboratory, the other in the field, showed that for males helping behavior decreased in a high-noise environment, and that social cues modulating such behavior under low-noise conditions were not responded to under high noise. The effects are attributed to the role of noise in reducing attention to peripheral stimuli.

351. McLean, E. K., & Tarnopolsky, A. Noise, discomfort
      and mental health: A review of the socio-medical
      implications of disturbance by noise.
      Psychological Medicine, 1977, 7, 19-62.

     A review of research on physical and mental effects of
noise, at both animal and human levels. Emphasis is placed
upon the important role of the meaning of noise, as well as
the unsatisfactory state of the evidence in regard to
effects of noise upon mental health.

352. Moran, S., & Loeb, M. Annoyance and behavioral
      aftereffects following interfering and
      noninterfering aircraft noise. Journal of Applied
      Psychology, 1977, 26, 719-726.

     Two separate experiments failed to confirm the type of
aftereffects found by Glass and Singer (#288) following
exposure to intermittent or continuous noise. Subjects
reported greater annoyance when working on a listening task
which the noise had an interfering effect than on a routine
task where there was no interference from noise.

353. Stave, A. M. The effects of cockpit environment on
      long-term pilot performance. Human Factors, 1977,
      19, 503-514.

     Pilot performance in a helicopter simulator improved
with increases in environmental stress in the form of
ambient noise levels and vibration, in spite of reports of
fatigue over 8-hour periods, and intermittent momentary
lapses in performance.

354. Ward, L. M., & Suedfeld, P. Human response to highway
      noise. Environmental Research, 1973, 6, 306-326.

Recordings of highway noise were used in a laboratory study of the effects of noise on cognitive performance, and in a five-day study of students living in a dormitory continuously exposed to the highway sounds. Results indicated no short-term effects on task performance, but diverse effects from the five-day exposure on individual and social behavior.

355. Weinstein, N. D. Effect of noise on intellectual performance. Journal of Applied Psychology, 1974, 59, 548-554.

A laboratory study, reporting differential effects of intermittent background teletype noise on proofreading tasks. Detection of grammatical errors was impaired, while detection of spelling errors was not; content recall was also unaffected. Data are presented on the manner in which subjects coped with the noise, and its effects on the pacing of their work.

356. Weinstein, N. D. Human evaluations of environmental noise. In #238 (Craik & Zube), pp. 229-252.

A review of methodological issues and empirical findings on effects of environmental noise on behavior and well-being, with particular emphasis on field studies.

357. Weinstein, N. D. Individual differences in reactions to noise: A longitudinal study in a college dormitory. Journal of Applied Psychology, 1978, 63, 458-466.

Prior to entering dormitories, freshmen were divided into noise sensitive and noise insensitive groups. Over a seven month period the sensitive group reported themselves more bothered by noise in the dorm; no change was found in the insensitive group. Relationships between noise-sensitivity and variables of personality and academic performance are reported.

358. Zimmer, J., & Brachulis-Raymond, J. Effects of
      distracting stimuli on complex information
      processing. Perceptual and Motor Skills, 1978,
      46, 791-794.

      Of three conditions of auditory distraction, popular
music, speech, and industrial noise, only the last showed
significant effects on performance in an information-
processing task.

359. Berning, A. B. Hearing loss in musicians: A
      preliminary investigation. Dissertation Abstracts
      International, 1977, 37-B, 3338.

360. DeJoy, D. M. Environmental noise: The effects of
      information overload and control over the work
      environment on performance during and after
      exposure. Dissertation Abstracts International,
      1978, 39-B, 637-638.

361. Dixon, P. J. The effects of noise on children's
      psychomotor, perceptual and cognitive performance.
      Dissertation Abstracts International, 1976, 37-A,
      3513-3514.

362. Ingram, E. Job performance in a high noise
      environment as related to selected individual
      differences. Dissertation Abstracts
      International, 1977, 37-B, 5870.

363. Siegel, J. M. Environmental noise level and the
      perception of social cues. Dissertation Abstracts
      International, 1978, 38-B, 4543.

See also: citations 241, 809, 962, 972, 982, 985, 1358,
1361, 1381.

e. Isolation and Extreme or Unusual Environments

364. Brownfeld, C. A. ISOLATION: CLINICAL AND EXPERIMENTAL
     APPROACHES. New York: Random House, 1965. 180 pp.

     A predominantly non-technical treatment of the clinical
and experimental literature dealing with effects of
isolation and sensory-deprivation, contains a balance of
laboratory and field-based research. Includes extensive
treatment of the phenomenon of "brain-washing" in prisoner
of war camps and its possible relation to sensory
deprivation.

365. Byrd, R. E. Alone. New York: Putnam's, 1938. 296
     pp.

     A perceptive and moving account of the author's
experience wintering over in isolation in a tiny outpost in
Antarctica. The impact of this isolation experience and of
the harsh physical environment to which the author was
exposed are vividly described.

366. Radloff, R., & Helmreich, R. GROUPS UNDER STRESS:
     PSYCHOLOGICAL RESEARCH IN SEALAB II. New York:
     Appleton-Century Crofts, 1968. 259 pp.

     A comprehensive investigation of the psychological
response of a group of 28 men who spent periods of 15 days
living and working in an underwater laboratory.

367. Rasmussen, J. (Ed.), MAN IN ISOLATION AND CONFINEMENT.
     Chicago: Aldine, 1973. ix, 330 pp.

An overview of both laboratory and field studies covering the psychological effects of isolation and confinement, including work on sensory deprivation, isolated groups living in the Antarctic, under water, and other conditions of geographic and social isolation.

368.   Zubek, J. P. (Ed.), SENSORY DEPRIVATION: FIFTEEN
        YEARS OF RESEARCH. New York: Appleton-Century
        Crofts, 1969. 522 pp.

An exhaustive review of the sensory-deprivation literature, with emphasis on laboratory findings, but including some reference to field research on effects of isolation, life in Antarctica, etc.

369.   Bevan, W. Behavior in unusual environments. In H.
        Helson and W. Bevan (Eds.). CONTEMPORARY
        APPROACHES TO PSYCHOLOGY. Princeton, N. J.: Van
        Nostrand, 1967. Pp. 385-418.

A discussion of alternative views of relations between the physical environment and behavior, along with a brief treatment of vigilance behavior and effects of special environmental conditions, particularly stimulus impoverishment, and weightlessness.

370.   Burvill, P. W. Mental health in isolated new mining
        towns in Australia. Australian and New Zealand
        Journal of Psychiatry, 1975, 9, 77-83.

A discussion of psychiatric problems faced by mine workers who have moved to geographically isolated new towns. These problems are ascribed to selective pre-existing traits characterizing such migrants, and aggravated by the isolation and other environmental and social attributes of the environments afforded by these towns.

371. Burvill, P. W., & Kidd, C. B. The two-town study:
A comparison of psychiatric illness in two
contrasting Western Australian Mining Towns.
Australian and New Zealand Journal of Psychiatry,
1975, 9, 85-92.

A comparison between psychiatric problems encountered
in an established town and a "boom" town, based primarily
on reports and records of general-practitioner physicians
in the two locations.

372. Butcher, J. N., & Ryan, M. Personality stability
and adjustment to an extreme environment. Journal
of Applied Psychology, 1974, 59, 107-109.

Compared personality traits of a group of Antarctic
explorers with those of a Control group, finding better
adjustment and higher achievement orientation in the former.
No major changes were observed during the course of the
winter spent in Antarctica.

373. Earls, J. H. Human adjustment to an exotic
environment: The nuclear submarine. Archives
of General Psychiatry. 1969, 20, 117-123.

Observations on conditions of life for the crew of a
nuclear submarine on a submerged two-month long patrol.
Emphasizes problems of sensory deprivation, isolation and
confinement, and the changes in the crew's response to
these conditions over the two months.

374. Gunderson, E. K. E. Emotional symptoms in extremely
isolated groups. Archives of General Psychiatry,
1963, 9, 362-368.

A study of the somatic and affective disturbances
encountered among a group of men who spent a winter in
different stations on the Antarctic continent.

375. Gunderson, E. K. E. Mental health problems in
        Antarctica. Archives of Environmental Health,
        1968, 17, 558-564.

A further report of emotional symptoms reported by Navy
personnel who wintered in Antaractica.

376. Haggard, E. A., As, A., & Borgen, C. M. Social
        isolates and urbanites in perceptual isolation.
        Journal of Abnormal Psychology, 1970, 76, 1-9.

Adverse effects from a sensory-deprivation experience
were found to be lower in a group of hermits in Norway, than
in a group of urban students.

377. Hollos, M., & Cowan, P. A. Social isolation and
        cognitive development: Logical operations and
        role-taking abilities in three Norwegian social
        settings. Child Development, 1973, 44, 630-641.

A comparison of 7- to 9-year old children growing up in
remote, isolated farms with those living in villages and
small towns, in terms of performance on diverse cognitive
and intellectual tasks. Evidence of deficit related to the
effects of isolation was limited to performance on social-
role taking tasks.

See also Hollos, M. Logical operations and role-taking
abilities in two cultures: Norway and Hungary. Child
Development, 1975, 46, 638-649.

378. Knapp, R. J., & Capel, W. C. Stress in the deep: A
        study of undersea divers in controlled dangerous
        situations. Journal of Applied Psychology, 1976,
        61, 507-512.

Verbal measures of anxiety were obtained from divers at different stages preceding, during and following a five-or ten-day period of underwater activity. Those with most previous diving experience showed highest levels of anxiety.

379. Nelson, P. D. Psychological aspects of Antarctic living. Military Medicine, 1965, 130, 485-489.

A descriptive account of the conditions of life in Antarctic Naval stations and the manner in which the men living there adapt to and cope with their environment, conditions of work and interpersonal stresses.

380. Popkin, M. K. et al. Novel behaviors in an extreme environment. American Journal of Psychiatry, 1974, 131, 651-654.

Studied incidences of two atypical forms of behavior, "staring" and "drifting," in a group of men wintering over at the South Pole, as possible indices of clinical depression.

381. Radloff, R. W. Research on life and work in undersea habitats. In W. M. Smith (Ed.), BEHAVIOR, DESIGN AND POLICY OF HUMAN HABITATS. Green Bay, Wisc.: University of Wisconsin-Green Bay, 1972. Pp. 33-54.

A descriptive and empirical account of the response of the Tektite underwater laboratory crews to life in such an environment, with emphasis on both individual and interpersonal aspects of behavior.

382. Ruff, G. E. Adaptation under extreme environmental conditions. In S. Klausner (Ed.), Society and its physical environment. Annals of the American Academy of Political and Social Science, 1970, 389, 19-26.

A discussion of mechanisms of biological and
psychological adaptation to conditions of environmental
stress, emphasizing the role of compensatory and
equilibrium-restoring processes, and of phenomena of
overcompensation in reactions to environmental extremes.

383. Savishinsky, J. S. Mobility as an aspect of stress in
      an Arctic community. American Anthropologist,
      1971, 73, 604-618.

Discusses patterns of short- and long-term mobility
among natives in fur-trading communities in the Canadian
North, as adjustive mechanisms to cope with stresses of
isolation, boredom and interpersonal tensions.

384. Sherman, M., & Key, C. B. The intelligence of
      isolated mountain children. Child Development,
      1932, 3, 279-290.

Representative of a number of studies carried out on the
intellectual development of children living in isolated
Appalachian mountain communities in the 1930's, this study
points to progressively increasing deficits in IQ with age,
as well as particularly severe retardation in the area of
verbal and language skills.

385. Smith, W. M. Observations over the lifetime of a
      small isolated group: Structure, danger, boredom
      and vision. Psychological Reports, 1966, 19,
      475-514.

A primarily descriptive account of the behavior and
adjustment of a small group of scientists who spent a summer
in exploration and research in Antarctica. Covers
perceptual, motivational and interpersonal aspects of their
experience.

386. Taylor, L. J., & Skanes, G. R. Psycholinguistic
        abilities of children in isolated communities of
        Labrador. Canadian Journal of Behavioral Science,
        1975, 7, 30-39.

     Presents intelligence-test profiles of white and Eskimo
children in Kindergarten and Grade 1 in several Labrador
communities. Evidence of retardation was greatest for
measures of verbal development.

387. Wright, M. Sisler, G., & Chylinski, J. A. Personality
        factors in the selection of civilians for isolated
        northern stations. Journal of Applied Psychology,
        1963, 47, 24-29.

     An investigation of personality characteristics
associated with favorable adjustment to a one-year period of
duty in an isolated Northern-Canadian post.

See also: citations 115, 461, 1049, 1059.

5. SOCIO-SPATIAL PROCESSES

The research covered in this chapter represents undoubtedly the most flourishing area within environmental psychology. The related issues of crowding, personal space, territoriality and privacy have all been active foci for research, with crowding the most popular by far of any topic in the field. This high level of research activity can be partially explained by the fact that these topics in many ways represent an outgrowth of social psychology, that is, they focus on the study of interpersonal and group relations across (empty) space. Indeed some (e.g. Epstein, see #63) have questioned whether this area of research properly belongs as a part of environmental psychology at all. However, given the central role this area occupies among those who consider themselves environmental psychologists, the answer to this question is apparent on purely pragmatic grounds.

Because of the voluminous literature on all of these topics, transcending as they do the boundary between psychology and sociology, it has not been possible to provide more than a sampling of the work actually available on each. Nevertheless, an attempt has been made to include all of the major literature reviews, bibliographies, and monograph titles. An emphasis has also been placed on inclusion of field studies which in many instances bring out more directly the environmental relevance of these topics than does much of the laboratory research. At the same time, an effort has been made to include a reasonably representative sample of the latter as well. With regard to territoriality and crowding, the material included is restricted almost exclusively to humans; in the case of the work on territoriality in particular, this represents a notable limitation of our coverage.

It should further be noted that classification of items into one of these four topics was not always easy. There are a few works that cover the field as a whole, or at least deal with it in a generic fashion; these have been placed in a short section of "General Works" which will be found at the head of this chapter. Finally, the distinction between

personal space and territoriality, though conceptually
tenable, often becomes blurred when applied to particular
situations; in general we have allowed the authors'
statements of their problems, or terms chosen in titles, to
determine the assignment of their work to one or another of
the categories. In a few cases, where a paper was clearly
relevant to both, they have been appropriately cross-
referenced.

a. General Works

388.   Altman, I. ENVIRONMENT AND SOCIAL BEHAVIOR: PRIVACY,
        PERSONAL SPACE, TERRITORIALITY, CROWDING.
        Monterey, California: Brooks-Cole, 1975. xi, 256
        pp.

    Develops an integrative model of socio-spatial behavior
based upon the regulation of interpersonal boundaries. The
central concept is privacy, viewed as the achievement of
desired levels of social interaction. Extensive
bibliography.

389.   Altman, I., & Chemers, M. M. CULTURE AND ENVIRONMENT.
        Monterey, CA: Brooks/Cole, 1980. xvi, 337 pp.

    An overview of relationships between culture and
response to the environment, with special reference to
environmental cognition, proxemics (privacy, personal space,
territoriality), residential arrangements and forms of
communal settlement.

390.   Ashcroft, N. & Scheflen, A. E. PEOPLE SPACE: THE
        MAKING AND BREAKING OF HUMAN BOUNDARIES. New
        York: Anchor, 1976, 185 pp.

    A non-technical treatment of diverse aspects of human
use of space, drawing on scientific literature, historical
material and incidental observation. (Paperback)

391. Esser, A. H. (Ed.) BEHAVIOR AND ENVIRONMENT: THE USE
      OF SPACE BY ANIMALS AND MEN. New York: Plenum,
      1979. xvi, 411 pp.

     Proceedings of an American Association for the
Advancement of Science Conference, including sections on
territoriality and dominance, space and contact behavior,
density and crowding, orientation and communication, and
communal behavior.

392. Hall, E. T. THE HIDDEN DIMENSION. New York:
      Doubleday, 1966. xii, 217 pp.

     A general treatment of "proxemics," people's response to
interpersonal distance and their use of space to control
interpersonal interaction. Offers a four-zone
differentiation of interpersonal distance, labeled
"intimate," "personal," "social," and "public," according to
the dominant function for each. Cultural differences in
interpersonal space and distance are emphasized.

392a. Pastalan, L. A., & Carson, D. H. SPATIAL BEHAVIOR OF
       OLDER PEOPLE. Ann Arbor, Michigan: Institute of
       Gerontology, University of Michigan - Wayne State
       University, 1970. 228 pp.

     A set of papers on territoriality, privacy, personal
space and other aspects of the ecology and use of space of
the elderly.

393. Unruh, D. R. Space and Environment: An Annotated
      Bibliography. Monticello, Ill.: Council of
      Planning Librarians, 1976. (Exchange
      Bibliographies #954), 51 pp.

     Includes material on housing, institutional space, space
and the elderly, urban design and form, etc.

394. Baldassare, M. Human spatial behavior. Annual Review
       of Sociology, 1978, 4, 29-56.

       A sociologist's overview of research and theory on
crowding, personal space and related problems.

395. Buttimer, A. Social space in interdisciplinary
              perspective. Geographical Review, 1969, 59,
              417-426.

       A review of sociologists' analyses of "social space,"
and of the role of spatial and geographic factors in socio-
spatial differentiation, with particular attention to the
thought of Chombart de Lauwe and Sorre.

396. Willems, E. P., & Campbell, D. E. One path through
              the cafeteria. Environment and Behavior, 1976, 8,
              125-140.

       A broad critical view of research and theory on privacy,
territoriality and crowding as represented by the papers of
Edney (#510), Stokols (#465), and others appearing in the
same issue. Focuses on logical and methodological problems
pervading this area of research, especially on the cognitive
and attitudinal emphasis prevalent in this literature, at
the expense of a consideration of overt behavior in relation
to its setting.

See also: citations 186, 273, 277.

b. Crowding

397. Booth, A. (Ed.) Human Crowding. Sociological
              Symposium, 1975, 14, 1-144.

Includes eight papers dealing with varied facets of crowding, from animal research and physiological effects to effects on aggression, neighborhood attitudes, family life and social pathology.

398. Booth, A.   URBAN CROWDING AND ITS CONSEQUENCES.   New
        York: Praeger, 1976.   xi, 139 pp.

An account of a large-scale field study of the relationships between residential crowding and aspects of individual and social behavior, and health and well-being. The findings disclose few consistent relationships and little evidence on adverse effects.   Occasional low-level relationships were found for household crowding, but rarely for neighborhood density.

399. Freedman, J. L.   CROWDING AND BEHAVIOR.   New York:
        Viking Press, 1975.   177 pp.

Compares varying definitions of density with quality of life measures.   Issues including crime, housing, and personal space are discussed with particular reference to urban settings.   Argues that density may have either positive or negative effects on well-being; its main role is in intensifying the expression of other variables.

400. Gurkaynak, M. R., & Lecompte, W. A. (Eds.) HUMAN
        CONSEQUENCES OF CROWDING.   New York: Plenum,
        1979.   331 pp.

Proceedings of a NATO Conference held in Turkey in 1977. Contains contributions by major researchers on crowding in the U.S.A. and abroad, in four categories: conceptual dimensions of crowding, experimental studies of consequences of crowding, multivariate studies of perceived crowding, and special topics.

401. Insel, P. M., & Lindgren, H. C. TOO CLOSE FOR
        COMFORT: THE PSYCHOLOGY OF CROWDING. Englewood
        Cliffs, N. J.: Prentice-Hall, 1978. 180 pp.

    A brief review of crowding, written at a non-technical
level and addressed to the introductory-level student or
layperson.

402. Living at high densities. Ekistics, 1965, 20, 182-219.

    A set of brief discussions of various aspects of density
related problems affecting human habitats, including
biological, psychological, social and administrative.
Includes papers by the geneticist Dobzhansky, the biologist
Waddington, the anthropologists E. T. Hall and M. Mead, and
the planners Giedion and Doxiadis.

403. Sobal, J. Density and crowding of man. Monticello,
        Ill.: Vance Bibliographies. (Architecture Series
        A-44) 59 pp.

    A comprehensive but unannotated bibliography on crowding
and related subjects, including both experimental and field
studies, and material from psychology, sociology and natural
recreation. The latter deals with problems of congestion
and social carrying-capacity in natural recreation areas.

404. Aiello, J. R., Epstein, Y. M., & Karlin, R. A.
        Effects of crowding on electrodermal activity.
        Sociological Symposium, 1975, 14, 43-58.

    Traces the course of changes in physiological arousal
during a 30-minute laboratory experience of crowding,
showing evidence that crowding results in increased arousal.

405. Aiello, J. R., Nicosia, G., & Thompson, D. E. Physiological, social and behavioral consequences of crowding on children and adolescents. Child Development, 1979, 50, 195-202.

   A laboratory study of effects of crowding on children aged 9 to 17, showing both physiological and behavioral effects, specifically increases in competitiveness and arousal.

406. Baum, A., Aiello, J. R., & Calesnick, L. E. Crowding and personal control: Social density and the development of learned helplessness. Journal of Personality and Social Psychology, 1978, 36, 1000-1011.

   Feelings of helplessness and loss of control, related to conditions of density and architectural design in a college dormitory, were studied through a combination of survey and experimental techniques (through the use of a Prisoner's Dilemma game). Helplessness was found to be an end-result of sustained unsuccessful attempts to establish individual control.

407. Baum, A., & Davis, G. E. Spatial and social aspects of crowding perception. Environment and Behavior, 1976, 8, 527-544.

   Studied the mediating effects of physical design features including room size, visual complexity and setting orientation on subjects' appraisals of crowding thresholds and available space in a scale model of constant size.

408. Baum, A., & Greenberg C. I. Waiting for a crowd: The behavioral and perceptual effects of anticipated crowding. Journal of Personality and Social Psychology, 1975, 32, 671-679.

Experimentally investigated the influence of expectations of increased density on subjects' behavior and evaluative ratings. Effects of anticipation of crowding were found on choice of seating location, experience of discomfort, ratings of the room and of others.

409. Baum, A., Harpin, R. E., & Valins, S. The role of group phenomena in the experience of crowding. Environment and Behavior, 1975, 7, 185-198.

A field test of the hypothesis that membership in social groups mitigates the stress of high density settings. Findings indicated that high density inhibited the formation of social groups but that once formed, the groups did help to mitigate the experience of crowding.

410. Baum, A. & Koman, S. Differential response to anticipated crowding: Psychological effects of social and spatial density. Journal of Personality and Social Psychology, 1976, 34, 526-536.

An experimental comparison of behavioral and attitudinal measures associated with anticipation of too many people (social density) vs. too little space (spatial density).

411. Baum, A., & Valins, S. Architectural mediation of residential density and control: Crowding and the regulation of social contact. Advances in Experimental Social Psychology 1979, 12, 131-175.

A review and analysis of crowding, in residential settings, as related to architectural features. Effects of anticipation of crowding, control over crowding, and phenomena of reactance and learned helplessness are discussed.

412. Baxter, J. C., & Deanovich, B. F. Anxiety-arousing
      effects of inappropriate crowding. Journal of
      Consulting and Clinical Psychology, 1970, 35,
      174-178.

      Anxiety levels of subjects experiencing experimentally
induced spatial invasion were assessed by their non-verbal
behavior and by the level of anxiety they attributed to
others in a projective test.

413. Bickman, L. et al. Dormitory density and helping
      behavior. Environment and Behavior, 1973, 5,
      465-490.

      Altruistic behavior in dormitories of varying densities
was studied using a questionnaire survey of attitudes as
well as unobtrusive observation. Cooperativeness,
friendliness and feelings of trust, as well as disposition
to engage in helping behavior were all found to be inversely
related to density.

414. Booth, A., & Johnson, D. R. The effects of crowding
      on child health and development. American
      Behavioral Scientist, 1975, 18, 736-749.

      Compared indices of density with measures of health,
development, and school performance. Congested households
were associated with small adverse effects on physical and
intellectual development.

415. Chandler, M. J., Koch, D., and Paget, K. F.
      Developmental changes in the response of children
      to conditions of crowding and congestion. In H.
      McGurk (Ed.), ECOLOGICAL FACTORS IN HUMAN
      DEVELOPMENT. Amsterdam: North-Holland, 1977.
      Pp. 111-123.

Under varying environmental conditions groups of nine
school-age children each were required to arrange themselves
in order of height. Results showed the physiological and
behavioral evidence of crowding to be a joint function of
the child's level of cognitive development and the
situational complexity of the task.

416. Choi, S. C., Mirjafari, A., Weaver, H. B. The concept
      of crowding: A critical review and proposal of an
      alternative approach. Environment and Behavior,
      1976, 8, 345-362.

Presents a model of crowding in terms of factors
affecting crowding (density, social, physical-environmental
and personal factors), and modes of adaptation to crowding
(behavioral cognitive, physiological).

417. Choldin, H. M. Urban density and pathology. Annual
      Review of Sociology, 1978, 4, 91-113.

A review of the literature, concluding that the
frequently hypothesized relationship between urban density
and indices of individual and social pathology has not been
satisfactorily demonstrated. Argues for greater attention
to social relations within households and their interaction
with density and other housing features.

418. Cohen, J. L., Sladen, B., & Bennett, B. The effects
      of situational variables on judgments of crowding.
      Sociometry, 1975, 38, 273-281.

Using the Desor technique (see #422), perceptions of
crowding in a simulated room were found to vary as a
function of type of activity (interactive vs. independent;
work vs. recreational) and according to whether the
occupants were acquainted or strangers.

419.   Cozby, P. G.   Effects of density, activity and
           personality on environmental preferences.   Journal
           of Research in Personality, 1973, 7, 45-60.

       Diverse situational variables (e.g., type of activity)
and experiential and personality measures were related to
personal space and density preferences.

420.   D'Atri, D. A.   Psychophysical responses to crowding.
           Environment and Behavior, 1975, 7, 237-252.

       Reviews the work on crowding with special attention to
the relation of density and elevation of blood pressure.
Presents findings of a study of relationships between blood
pressure levels and various forms of housing in correctional
institutions.

421.   Dean, L. M., Pugh, W. M., & Gunderson, E.   Spatial and
           perceptual components of crowding: Effects on
           health and satisfaction.   Environment and
           Behavior, 1975, 7, 225-236.

       Compared indices of crowding with attitudinal and
behavioral measures including job satisfaction, illness and
accident rates, and organizational climate of crew members
on thirteen Navy vessels.

422.   Desor, J. A.   Toward a psychological theory of
           crowding.   Psychological Review, 1972, 21, 79-83.

       A series of three studies utilizing a simulation method
(by presenting subjects with models of rooms, with
instructions to fill them with miniature human figures until
the room appeared "crowded") to study the influence of
particular architectural variables (including partitions and
number of doors), on perceived crowding.   The latter is
conceptualized as excess social stimulation.

423.   Edwards, J. N., & Booth, A. Crowding and human sexual
       behavior. Social Forces, 1977, 55, 791-808.

Found few relationships between density of urban
residences and patterns of sexual behavior or sex-related
disturbances.

424.   Epstein, Y. M., & Karlin, R. A. Effects of acute
       experimental crowding. Journal of Applied Social
       Psychology, 1975, 5, 34-53.

Demonstrated effects of experimentally induced crowding
on arousal, but found contrasting overt manifestations in
the two sexes.

425.   Esser, A. H. A biological perspective on crowding.
       In #14 (Wohlwill & Carson), pp. 15-28.

An interpretation of the subjective experience of
crowding in terms of physiological and neurological
processes.

426.   Evans, G. W. Behavioral and physiological
       consequences of crowding in humans. Journal of
       Applied Social Psychology, 1979, 9, 27-46.

Reports a variety of physiological and behavioral
indices of the arousing and stressful effects produced by
experimentally induced crowding. Performance decrements
were also found for complex, but not simple tasks.

427.   Freedman, J., Levy, A., Buchanan, R., & Price, J.
       Crowding and human aggressiveness. Journal of
       Experimental Social Psychology, 1972, 8, 528-548.

Aggressiveness, operationalized as competitiveness and severity of judgment of others, was related to population density in large and small rooms. Findings showed that within same-sex groups, males responded more positively in larger rooms and females more positively in smaller rooms.

428. Galle, O., Gove, W., & McPherson, J. Population
      density and social pathology: What are the
      relationships for man? Science, 1972, 176, 23-30.

This paper, a landmark in the literature of survey research on the effects of residential density, differentiates among different aspects of density: number of persons per room and per dwelling, number of housing units per structure, and number of structures per acre. Some evidence for relationships between particular indices of density and of social pathology emerge when confounding roles of socio-economic and ethnic variables are partialled out.

429. Glassman, J. B., Burkhart, B. Z., Grant, R. D., &
      Vallery, G. G. Density, expectation, and extended
      task performance. Environment and Behavior, 1978,
      10, 299-316.

A field study of the effects of density on behavior, comparing students residing in high and low density dormitories. Results point to negative effects of high density on both performance (grade-point average) and satisfaction.

430. Greenberg, C. I., & Baum, A. Compensatory response to
      anticipated densities. Journal of Applied Social
      Psychology, 1979, 9, 1-12.

Studied the role of expectations of crowding on the nature of preparatory coping mechanisms elicited, as well as responses to conditions of crowding either congruent with or disconfirmed by the social situation into which subjects were placed.

431.   Greenberg, C. I., & Firestone, I. G.   Compensatory
        responses to crowding: Effects of personal space
        intrusion and privacy reduction. Journal of
        Personality and Social Psychology, 1977, 35,
        637-644.

       This experiment manipulated spatial invasions, social
surveillance by strangers, and restriction of withdrawal
behavior by seat location. Observations of overt coping
behaviors and self-reports of crowding were used as measures
of crowding stress. Both intrusion and surveillance
produced significant and additive effects.

432.   Griffitt, W., & Veitch, R.   Hot and crowded:
        Influences of population density and temperature
        on interpersonal affective behavior. Journal of
        Personality and Social Psychology, 1971, 17,
        92-98.

       Ratings of subjects' mood, and their evaluative
judgments of the experimental room were compared for
settings varying in temperature and density.

433.   Gruchow, H. W.   Socialization and the human
        physiologic response to crowding. American
        Journal of Public Health, 1977, 67, 455-459.

       Suggests that the effects of prolonged residential
crowding on physiological stress response may be mediated by
the individual's socialization in childhood in the home.
This conclusion is derived from the finding that the
relationship between crowding and stress was positive for
individuals without younger siblings, and with few older
siblings, but negative for those with younger siblings.

433a. Hutt, C., & Vaizey, M. J.   Differential effects of
        group density on social behavior. Nature, 1966,
        209, 1371-1372.

One of the classics in the literature on crowding, demonstrating effects of increasing density on the aggressiveness of nursery-school children and a converse decrease in social interaction. Results for autistic and brain-damaged children were somewhat different.

434. Kutner, D. H., Jr. Overcrowding: Human responses to density and visual exposure. Human Relations, 1973, 26, 31-50.

Assessed stress responses by experimentally manipulating the number of people, interpersonal distance, and amount of visual access to the subject.

435. Lawrence, J. E. Science and sentiment: Overview of research on crowding and human behavior. Psychological Bulletin, 1974, 81, 712-720.

A brief review and critique of research on relationships between density and behavioral impairment and pathology, with emphasis on demographic and field studies.

436. Loo, C. M. The effects of spatial density on the social behavior of children. Journal of Applied Social Psychology, 1972, 2, 372-381.

Experimental observations of various types of social interactions by 4 and 5-year olds in free-play settings were made. Less aggressive behavior and reduced social interaction were found under high density. Sex differences were noted.

437. Loo, C. M. Important issues in researching the effects of crowding on humans. Representative Research in Social Psychology, 1973, 4, 219-226.

A discussion, outlining various methods, definitions,
and approaches used in the study of crowding and density,
together with implications for their use.

438. Loo, C. M. The psychological study of crowding:
        Historical roots and conceptual development.
        American Behavioral Scientist, 1975, 18, 826-842.

Traces the development of and influences on environment
and behavior research, with emphasis on crowding studies.
Discusses various models of density and crowding.

439. Loo, C. M. Beyond the effects of crowding:
        Situational and individual differences. In #13
        (Stokols), pp. 153-168.

Considers effects of crowding in relation to
environmental factors, activity engaged in, territoriality,
and individual differences. Concludes with a schematic
model for the analysis of crowding, differentiating between
social and spatial aspects.

440. Loo, C. M. Behavior problem indices: The
        differential effects of spatial density on low and
        high scorers. Environment and Behavior, 1978, 10,
        489-510.

Demonstrated interactions between density and certain
personality measures such as hyperactivity, anxiety and
behavior disturbance in influencing various behavioral
measures obtained from 5 year old children.

441. Mackintosh, E., West, S., & Saegert, S. Two studies
        of crowding in urban public spaces. Environment
        and Behavior, 1975, 7, 159-184.

Reports the results of field-experiments of the effects
of congestion on shoppers in a department store and on users
of a railroad terminal. The negative effects on task
performance, cognitive mapping, and affect were interpreted
in terms of the increased complexity and uncertainty
associated with crowded environments.

442. McCain, G., Cox, V. C., & Paulus, P. B. The
relationship between illness complaints and degree
of crowding in a prison environment. Environment
and Behavior, 1976, 8, 283-290.

Compared potentially stress related illness complaints
of inmates in the high and low density conditions of a
prison and county jail. Such complaints were found to be
more frequent under high density conditions, a finding
attributed to the mediating role of psychological stress
from crowding.

443. McClelland, L., & Auslander, N. Perceptions of
crowding and pleasantness in public settings.
Environment and Behavior, 1978, 10, 535-554.

Studied the relationship between perceived crowding and
pleasantness, using college students' ratings of a large
variety of environmental settings presented via slides.
Substantial correlations were limited to work and shopping
environments; for other environments the relationship was
modulated by the role of other variables.

444. McGrew, P. L. Social and spatial density effects on
spacing behavior in preschool children. Journal
of Child Psychology and Psychiatry, 1970, 11,
197-205.

Compares the effects on socio-spatial behavior of social
density (varying number of children in a room of constant
size) with those of spatial density (varying room size for a
constant number of children).

445. McPherson, J. M. Population density and social
      pathology: A re-examination. Sociological
      Symposium, 1975, 14, 77-90.

      A re-analysis of the data of Galle, Gove and McPherson
(#428) by means of cross-lagged panel analysis (as compared
to the cross-sectional correlations of the original report)
leads to the conclusion that clear effects of crowding on
social pathology have not been demonstrated. The model of
temporally directional effects underlying this analysis is
contrasted to the typical atemporal one implicit in ordinary
correlational analysis.

446. Mitchell, R. E. Some social implications of high
      density housing. American Sociological Review,
      1971, 36, 18-29.

      Studied the influence of density (number of families per
dwelling) on individual and marital satisfaction. Although
congestion was noted as a potentally stressful factor, very
little evidence pointed to the influence of density per se.

447. Murray, R. The influence of crowding on children's
      behavior. In #48 (Canter & Lee), pp. 112-117.

      Showed a relationship between amount of aggressiveness
of school-age children (as rated by their peers) and two
indices of crowding in the home. Also found was a
relationship between crowding and certain other personality
measures, which differed by sex.

448. Paulus, P. B. et al. Density does affect task
      performance. Journal of Personality and Social
      Psychology, 1976, 34, 248-253.

      Reports a trio of experiments manipulating group size,
personal space, and room size on performance of a maze task.
Sex differences were noted.

449. Paulus, P., Cox, V., McCain, G., & Chandler, J.   Some
      effects of crowding in a prison environment.
      Journal of Applied Social Psychology, 1975, 5,
      86-91.

      A comparison of prison inmates housed under conditions
of social crowding with others living under less crowded
conditions. The crowded inmates showed higher sensitivity
to crowding (based on the Desor test of distances at which
doll figures are placed to indicate crowding) as well as
more negative evaluations of the environment.

450. Rapoport, A. Toward a redefinition of density.
      Environment and Behavior, 1975, 7, 133-158.

      A theoretical argument for the inclusion of a subjective
component in definitions of density as well the experience
of crowding. A distinction is drawn between the number of
people in a space (physical density) and perceived density,
which allows for individual and cultural variation.
Perceived density is then compared to norms for its settings
to produce affective density, of which crowding is one
component.

451. Rawls, J. R., Trego, R. E., McGaffey, C. N., & Rawls,
      D. J. Personal space as a predictor of
      performance under close working conditions.
      Journal of Social Psychology, 1972, 86, 261-267.

      Reports a pair of experiments probing at three levels of
density the psychomotor and arithmetic performance of males
testing high and low on personal space measures.

452. Rodin, J. Density, perceived choice, and response to
      controllable, and uncontrollable outcomes.
      Journal of Experimental Social Psychology, 1976,
      12, 564-578.

Findings showed that children from high density
environments were less likely to try to control the
experimental situation when the opportunity arose, and were
significantly poorer at mastering soluble problems preceded
by unsoluble ones than were children from lower residential
densities.

453. Rohe, W., & Patterson, A. H. The effects of varied
      levels of resources and density on behavior in a
      daycare center. In #38 (EDRA-5), vol.12, pp.
      161-171.

A field experiment investigating children's social
behavior in relation to experimentally varied spatial
density (room size) and number of toys available. The role
of resources in mitigating the effects of increased density
is discussed.

454. Ross, M., Layton, B., Erickson, B., & Schopler, J.
      Affect, facial regard, and reactions to crowding.
      Journal of Personality and Social Psychology,
      1973, 28, 69-76.

Subjects' ratings of themselves and others, together
with frequency of facial regard were analyzed at three
levels of density. Significant sex differences were noted.

455. Saegert, S. Crowding: Cognitive overload and
      behavioral constraint. In #37 (EDRA #4), Vol. 2,
      pp. 254-260.

Reports a field experiment exploring cognitive and
behavioral aspects of two physical components of density:
the number of people per space, and the amount of space per
person.

456. Schiffenbauer, A. I. et al. The relationship between
      density and crowding: Some architectural
      modifiers. Environment and Behavior, 1977, 9,
      3-14.

      Explored the relationship of levels of illumination,
expansiveness of view from room windows, and varying amount
of usable space on evaluations of room size and crowdedness.
Perceived room size and judged levels of crowding were found
to be independent.

457. Schmidt, D. E., Goldman, R. D., & Feimer, N. R.
      Physical and psychological factors associated with
      perceptions of crowding: An analysis of
      subcultural differences. Journal of Applied
      Psychology, 1976, 61, 279-289.

      A field study analyzing characteristics associated with
varying levels of density by black, white, and Spanish
groups.

458. Schmidt, D. E., Goldman, R. D., & Feimer, N. R.
      Perceptions of crowding: Predicting at the
      residence, neighborhood and city levels.
      Environment and Behavior, 1979, 11, 105-130.

      Compared the role of physical indices of density with a
variety of psychological scales as predictors of crowding.
The former decreased, while the latter increased in their
predictive values, as geographic scale increased from
residence to neighborhood to city.

459. Schmidt, D. E., & Keating, J. P. Human crowding and
      personal control: An integration of the research.
      Psychological Bulletin, 1979, 86, 680-700.

      A selective, integrative review of the crowding
literature, leading to a conceptual model linking
situational density with loss of personal control.

460.   Schopler, J., & Stockdale, J. E.   An interference
       analysis of crowding. Environmental Psychology
       and Nonverbal Behavior, 1977, 1, 81-88.

A view of crowding as based in interference in goal-
directed behavior, as a function of spatial arrangements,
resource scarcity, task demands and personal expectations.
The model is applied to questionnaire data on the
residential satisfaction of students in two types of
dormitories.

461.   Smith, E. J., & Lawrence, J. E.   Alone and crowded:
       The effects of spatial restriction on measures of
       affect and simulation response. Personality and
       Social Psychology Bulletin, 1978, 4, 139-142.

The perceptions of crowding, ratings on scales of
affect, and selected behavioral measures of subjects
confined singly in chambers of two different sizes were
studied. The more spatially restricted chamber yielded
enhanced sense of crowding, and more negative affect.

462.   Sommer, R., & Becker, F. D.   Room density and user
       satisfaction. Environment and Behavior, 1971, 3,
       412-417.

Related the number of people per class in varying sized
classrooms with measures of satisfaction of both teachers
and students over a four-year period.

463.   Stockdale, J. E. Crowding: Determinants and effects.
       Advances in Experimental Social Psychology, 1978,
       11, 197-247.

A comprehensive, integrative review, which examines
conceptual and theoretical issues, and work on specific

factors that modulate effects of crowding, including group
size, membership and structure, task activity, resource
scarcity, environmental features, and differences related to
sex and personality.

464. Stokols, D. On the distinction between density and
     crowding: Some implications for future research.
     Psychological Review, 1972, 79, 275-277.

Differentiates between the physical condition of density
as numbers of people in a space and the stressful experience
of crowding, considered as a disparity between the supply
and desire for space. Outlines strategies for assessing the
experience of crowding.

465. Stokols, D. The experience of crowding in primary and
     secondary environments. Environment and Behavior,
     1976, 8, 49-86.

Critically reviews stimulus overload, behavioral
constraint, and ecological models of crowding. Presents an
alternative conception of crowding based on personal vs.
neutral thwartings and a differentiation of primary and
secondary environments.

466. Stokols, D., Rall, M., Pinner, B., & Schopler, J.
     Physical, social, and personal determinants of the
     perception of crowding. Environment and Behavior,
     1973, 5, 87-117.

Reports an experimental study of self-reported anxiety,
game enjoyment, aggressiveness, and comfort in relation to
varying levels of density produced by putting the same
number of people in different sized rooms. For several of
the behavioral measures important sex differences, as well
as interactions between sex and room size were found.

467.   Stokols, D., Smith, T., & Prostor, J. Partitioning
        and perceived crowding in a public space.
        American Behavioral Scientist, 1975, 18, 792-814.

     A field experiment designed to test the validity of
laboratory work which indicated a mitigating influence of
architectural features on the experience of crowding.
Laboratory findings were not confirmed.

468.   Sundstrom, E. An experimental study of crowding:
        Effects of room-size, intrusion, and goal-blocking
        on non-verbal behavior, self-disclosure, and self-
        reported stress. Journal of Personality and
        Social Psychology, 1975, 32, 645-654.

     Report of an experiment well described by its title.
Findings are discussed in light of a sequential,
interpersonal model of crowding.

469.   Valins, S., & Baum, A. Residential group size, social
        interaction, and crowding. Environment and
        Behavior, 1973, 5, 421-439.

     Compares the quality and quantity of social interactions
experienced by residents of high and low density
dormitories. Stress related implications of socially
overloaded environments are discussed.

470.   Welch, S., & Booth, A. The effect of crowding on
        aggression. Sociological Symposium, 1976, 14,
        105-127.

     A field study of urban families failed to show any
consistent relationship between residential density and
measures of aggression in every-day life (arguments, etc.)

471. Worchel, S., & Teddlie, C. The experience of
         crowding: A two-factor theory. Journal of
         Personality and Social Psychology, 1976, 34,
         30-40.

     Interaction distance, room density, and 'attribution
inhibitors' were manipulated in a test of the theory that
crowding is experienced as arousal due to violated personal
space, with the arousal being attributed to others in the
environment.

472. Zlutnick, S., & Altman, I. Crowding and human
         Behavior. In #14 (Wohlwill and Carson), pp.
         44-60. biographical-historical,

     A review and conceptualization of the effects of
crowding, including material on its physical, psychological
and social effects. Emphasizes the differentiation between
"inside" and "outside" density, as well as temporal duration
and environmental resources as modulators of crowding
effects.

473. Baldassare, M. G. Residential crowding in urban
         America. Dissertation Abstracts International,
         1977, 38-A, 1029-1030.

474. Clark, S. A. Territorial behavior as a function of
         changing environments. Dissertation Abstracts
         International, 1977, 37-B, 4111.

     (Involved studying response to relocation to a crowded
environment.)

475. Gochman, I. R. Causes of perceived crowding unrelated
         to density. Dissertation Abstracts International,
         1977, 37-B, 3675-3676.

476. Greenberg, C. I. Intimacy overload as a function of
     intrusion, surveillance, and behavioral
     restriction: An interpersonal distance
     equilibrium approach to the experience of
     crowding. Dissertation Abstracts International,
     1977, 37-B, 5879.

477. Spiro, H. R. The relationships of crowding to
     emotional and physical distress. Dissertation
     Abstract International, 1976, 37-B, 531-532.

See also: citations 328, 737, 757, 953, 1024, 1029, 1044,
1063, 1064, 1065, 1075, 1095, 1121, 1155, 1170, 1193, 1230,
1243, 1246, 1252, 1253, 1254, 1256, 1263, 1264, 1289, 1299,
1300, 1314, 1393.

c. Privacy

478. Margulis, S. T. (Ed.) Privacy as a behavioral
     phenomenon. Journal of Social Issues, 1977, 33,
     (3), 1-199.

     A special issue devoted to the topic of privacy, with
contributions concerning diverse psychological aspects, as
well as biological, cultural, legal and architectural
perspectives.

479. Pennock, J. R., & Chapman, J. W. (Eds.). PRIVACY.
     New York: Atherton Press, 1971. xx, 250 pp.

     A collection of essays providing perspectives on privacy
from diverse disciplines, including philosophy, sociology,
political science, anthropology and law.

480. Westin, A. F. PRIVACY AND FREEDOM. New York:
     Atheneum, 1967. 487 pp.

A classic exposition of the functions of privacy. Autonomy, emotional release, self-evaluation, and individual control over information communicated to others about the self are discussed.

481. Johnson, C. A. Privacy as personal control. In #38
      (EDRA 5), Vol. 6, pp. 83-95.

A theoretical paper which argues for consideration of privacy issues in terms of individual effectance. Outlines a four-step model of personal control and its potential contribution to privacy studies.

482. Kira, A. Privacy and the bathroom. In #29
      (Proshansky et al., 1970), pp. 268-275.

A discussion of social, economic, cultural, and individual factors influencing the relative degrees of privacy sought for personal hygiene functions.

483. Konyea, M. Privacy regulation in small and large
      groups. Group and Organization Studies, 1977, 2,
      324-335.

A discussion of the relationship between the satisfaction of individual privacy needs, considered as control over information about self transmitted to others, and seating arrangement in groups of varying size.

484. Laufer, R. S., Proshansky, H. M., & Wolfe, M. Some
      analytic dimensions of privacy. In #50 (Kuller)
      pp. 353-372.

Argues for a treatment of privacy in terms of component dimensions rather than global definitions. A set of eight

dimensions are proposed: self-ego, interaction,
biographical-historical control, ecological-cultural, task
orientation, ritual privacy and phenomenological.

485. Lawton, M. P., & Bader, J. Wish for privacy by young
        and old. Journal of Gerontology, 1970, 25, 48-54.

A survey of the elderly's preferences for private as
opposed to shared living accomodations. A variety of
factors regarding privacy preference are discussed and
analyzed including age, sex, institutionalization, and
socio-economic status.

486. Marshall, N. J. Privacy and environment. Human
        Ecology, 1972, 1, 93-110.

Explores the 'fit' of environments by comparisons of
past and present density levels with preferred levels of
privacy.

487. Parke, R. D., & Sawin, D. B. Children's privacy in
        the home: Developmental, ecological and child-
        rearing determinants. Environment and Behavior,
        1979, 11, 87-104.

A study of factors related to privacy behavior in
children between 2 and 17 years, stressing the interplay
between developmental and situational factors, as well as
other variables such as sex.

488. Pastalan, L. A. Privacy as a behavioral concept.
        Social Science, 1970, 45, 93-97.

Outlines four modes of privacy: solitude, intimacy,
anonymity and reserve. Develops a model of antecedent
factors, organismic factors, behavior states, and
environmental factors for each mode. Operationalizes
privacy in terms of control of information about the self.

489. Pastalan, L. A. Privacy preferences among relocated
        institutionalized elderly. In #38 (EDRA 5), Vol.
        6, pp. 73-82.

    Reports the results of privacy-preference scales
administered to relocated and nonrelocated elderly patients.
Counter to expectation, patients showed preference for low
levels of privacy. These results are discussed with
reference to the limited opportunities for privacy among
institutionalized elderly.

490. Proshansky, H. M., Ittelson, W. H., & Rivlin, L. G.
        Freedom of choice and behavior in a physical
        setting. In #30 (Proshansky et al., 1976), pp.
        170-181.

    Uses privacy as an organizing vehicle for discussions of
individuals' goal-directed use of the environment. Outlines
various types of privacy and other social functions of space
use including territoriality and crowding.

491. Roberts, J. M., & Gregor, T. Privacy: A Cultural
        view. In #479 (Pennock and Chapman), pp. 199-225.

    A comparative look at mechanisms for controlling
channels of communication within various cultures, and the
arrangements and use of space devised for this purpose. The
development of a sense of privacy as a product of cultural
development is also discussed.

492. Schwartz, B. The social psychology of privacy.
        American Journal of Sociology, 1968, 73, 741-752.

    Privacy is discussed as the reciprocal of social
interaction, reducing potential conflict, irritation, or
'dysfunctional knowledge'. Physical elements of space are

described in terms of rules and regulations for social
engagement and disengagement.

493. Sommer, R. The ecology of privacy. In #29
          (Proshansky et al., 1970), pp. 256-66.

      Uses a library setting to demonstrate how designed
environments influence social interaction. Reports a series
of experiments on the use of space in a setting designed to
discourage social interaction.

494. Wolfe, M. Childhood and privacy. In #10 (Altman &
          Wohlwill, 3), pp. 175-222.

      An examination from a developmental perspective of the
concept of privacy, and research on privacy in childhood
through the lifespan.

495. Wolfe M., & Laufer, R. The concept of privacy in
          childhood and adolescence. In #38 (EDRA 5, Vol.
          6), pp. 29-54.

      Addresses issues of human privacy from a developmental
perspective. Interviews with children and adolescents (ages
5-17) pointed to four major definitions of privacy:
aloneness, control of information, control of space, and
lack of interference from others.

496. Rauter, U. K. A multivariate analysis of the
          contribution of environmental and personality
          components to privacy satisfaction in a sample of
          mental hospital residents. Dissertation Abstracts
          International, 1978, 39-B, 2516-2517.

497. Smith, D. E. Privacy and environment: A field
          experiment. Dissertation Abstract International
          1978, 38-B, 3475.

See also: citations 721, 1067, 1070, 1105, 1109, 1276, 1357, 1359, 1466, 1468.

d. Territoriality and Personalization of Territory

498. Allen, P. R. B. An annotated bibliography of mostly obscure articles on human territorial behavior. Monticello, Ill.: Council of Planning Librarians, 1975. (Exchange Bibliographies #754) 17 pp.

Includes material on personal space and privacy.

499. Acheson, J. M. Territories of the lobstermen. Natural History, 1972, 4, 60-69.

An informal description of ways that New England lobstermen trollers have devised to divide the resource on which their livelihood depends. Also recounts the escalating hierarchy of tactics used to defend those territories against newcomers who may not recognize or respect the claims.

500. Acheson, J. M. The lobster fiefs: Economic and biological effects of territoriality in the Maine lobster industry. Human Ecology, 1975, 3, 183-207.

Similar to the preceding, but presented in a more systematic and formal fashion. Suggests a distinction between "nuckated" and "perimeter-defended" territories.

501.   Altman, I. Territorial behavior in humans: An
          analysis of the concept. In #392a (Pastalan &
          Carson), pp. 1-24.

   Reviews concepts of human territoriality with attention
to potential differences from the mechanisms of other
animals on which most research has been done. Develops a
broadly stated framework for definitions of territoriality
and outlines future directions for research.

502.   Altman, I., & Haythorn, W. W. The ecology of isolated
          groups. Behavioral Science, 1967, 12, 169-182.

   An investigation of territorial behavior and other modes
of behavioral adaptation observed in men isolated in dyads
in a small, minimally furnished and equipped suite. An
interesting feature is the use of personality inventories to
create dyads differing in the compatibility of their
personality syndromes.

503.   Becker, F. D. Study of spatial markers. Journal of
          Personality and Social Psychology, 1973, 26,
          439-445.

   Two experimental field studies investigated the
influence of room density and the presence of personal
belongings used to claim public space on seating
arrangements and duration of space use. The role of spatial
markers as aids to conflict avoidance was also tested.

504.   Becker, F. D., & Mayo C. Delineating personal
          distance and territoriality. Environment and
          Behavior, 1973, 3, 375-381.

   Used a field experiment to differentiate between use of
personal belongings to maintain comfortable interpersonal
spacing and the defense of a localized territory valued as
such by its claimant.

505. Blood, R. L., & Livant, W. P. The use of space within
      the cabin group. Journal of Social Issues, 1957,
      13(1), 47-53.

      Studied use and manipulation of space at a Fresh Air
camp by a group of maladjusted boys.

506. Davis, G., & Altman, I. Territories at the work-
      place: Theory into design guidelines. Man-
      Environment Systems, 1976, 6, 46-53.

      The process of translating theoretical concepts of
interpersonal boundary control such as privacy and
territoriality to design guidelines for the renovation of
office space is described. The development of a user's
manual for occupants of the renovated space is also
discussed.

507. DeLong, A. J. Dominance-territorial relations in a
      small group. Environment and Behavior, 1972, 2,
      170-191.

      Describes a case study of the relationship between group
leadership roles and seating patterns, based on observations
of a small seminar over an eight week period.

508. Edney, J. J. Property, possession, and performance:
      A field study of human territoriality. Journal of
      Applied Social Psychology, 1972, 2, 275-282.

      Probed the relationships among territorial markings
("Private Property" and "Keep Out" signs) and physical
barriers (i. e. fences) with length of residence in a
particular house, expected future residence, and speed of
response by residents to the arrival of a stranger at their
front door.

509. Edney, J. J. Human territoriality. Psychological
          Bulletin, 1974, 81, 959-975.

     A review of the literature, with major emphasis on work
at the human level. The empirical and theoretical
literature points to a diversity of definitions and of
methodological approaches to study in this field, and to a
lack of cohesiveness of this body of research.

510. Edney, J. J. Human territories: Comment on
          functional properties. Environment and Behavior,
          1976, 8, 31-48.

     Outlines the importance of territoriality in human
social behavior at individual, group, and community levels.
Argues for a definition of territoriality in humans
emphasizing consistent use rather than aggressive defense of
space.

511. Edney, J. J. Territoriality and control: A field
          experiment. Journal of Personality and Social
          Psychology, 1975, 31, 1108-1115.

     Compared the behavior of occupants of rooms in a college
dormitory with that of visitors to the same rooms, to obtain
evidence on effects of territoriality on experience of
control, privacy and space-related affect.

512. Edney, J. J., & Buda, M. A. Distinguishing
          territoriality and privacy: Two studies. Human
          Ecology, 1976, 4, 283-296.

     Reports a study of people's stated preferences for
settings with varying combinations of territoriality and
privacy when performing biological, work, recreational, and
"unconventional" activities. Also reports a field
experiment on the roles of privacy and territoriality in
evaluative judgements of rooms and perception of control.

513.   Edney, J. J., & Jordon-Edney, N. Territorial spacing
       on a beach. Sociometry, 1974, 37, 92-104.

       Investigated the relation of amount of space claimed in
field settings to number of people in the group, sex
differences, and density of other uses. Studied
longitudinal changes in amount of space claimed and use
estimates of crowding.

514.   Esser, A. H., Chamberlain, A. S., Chapple, E. D. &
       Kline, N. S. Territoriality of patients on a
       research ward. In J. Wortis (Ed.), RECENT
       ADVANCES IN BIOLOGICAL PSYCHIATRY, Vol. 7, New
       York: Plenum Press, 1965, pp. 37-44.

       Reports an investigation of the relationships of
territorial behaviors and dominance hierarchies in a group
setting. Used unobtrusive observation to study the
interactions of group status, verbal behavior, freedom of
movement, establishment and defense of territories, and
aggressive behaviors.

515.   Hall, E. T. Environmental communication. In #391
       (Esser), pp. 247-256.

       Contrasts the territorial behavior patterns of blacks
and whites. Describes zones of spatial control and the use
of territories in child rearing, social organization and
cohesion in black communities. Territorial patterns of
blacks are compared to the physical structure of high-rise,
public housing.

516.   Hansen, W. B., & Altman, I. Decorating personal
       places: A descriptive analysis. Environment and
       Behavior, 1976, 8, 491-504.

Modes of personalizing living spaces were analyzed in terms of the timing, content, and longitudinal changes of the modifications made.

517.   Ley, D., & Cybriwsky, R. The spatial ecology of stripped cars. Environment and Behavior, 1974, 6, 53-68.

Crime statistics were used to test the hypothesis that areas with higher territorial control (i. e.; those which were constantly surveyed and defended by users) experienced less vandalism.

518.   Patterson, A. H. Territorial behavior and fear of crime in the elderly home owner. Environmental Psychology and Nonverbal Behavior, 1978, 2, 131-144.

Explored the relationship between territorial markers used by urban elderly to personalize and defend their homes and the level of fear they expressed about crime against themselves or their property. Findings indicated that those with greater numbers of territorial markers expressed less anxiety about crime.

519.   Shaffer, D. R., & Sadowski, C. This table is mine: Respect for marked barroom tables as a function of gender of spatial marker and desirability of locale. Sociometry, 1975, 38, 408-419.

A field experiment studied the effectiveness of spatial markers representing a male owner vs. those representing a female owner in postponing or preventing use of the marked area. The influence of the space's attractiveness on the marker's ability to reserve it was also probed.

520.   Sommer, R. Spatial parameters in naturalistic social research. In #391 (Esser), pp. 281-290.

A broad-ranging discussion of the relationship of two major threads of spatial research: dominance (vs. subordinance) and territoriality. Territoriality is dimensionalized as the personalization of space and its defense. Examples are drawn from recreational, institutional, and academic settings.

521. Sommer, R., & Becker, F. D. Territorial defense and the good neighbor. Journal of Personality and Social Psychology, 1969, 11, 85-92.

A series of seven studies explored strategies for designation and defense of space in public. The effectiveness of various types of markers and their use in signaling active defense of area or passive retreat were studied under several levels of room density. The role of neighbors in supporting spatial claims was also investigated.

522. Stea, D. Home range and use of space. In #392a (Pastalan & Carson), pp. 138-147.

Uses social, physical, spatial, and cognitive components to define the concept of home range and differentiate it from that of territoriality. Traces the development and evolution of home range patterns and correlated patterns of environmental imagery throughout the life cycle.

523. Taylor, R. B., & Stough, R. R. Territorial cognition: Assessing Altman's typology. Journal of Personality and Social Psychology, 1978, 36, 418-423.

Provides empirical confirmation for the differentiation among different types of human territories i.e., primary, secondary and public, proposed by Altman, and relates these to differences between urban vs. suburban residents.

524. VanDenBerghe, P. L. Territorial behavior in a natural
     human group. Social Science Information, 1977,
     16, 419-430.

A field study of manifestations of territoriality on the
part of persons' holding plots in an outdoor recreation area
through membership in a recreational club, and their family
members. Defense of territory is found at three levels:
within the family (i.e., specific areas of a cabin), against
fellow club members owning other plots, and against
outsiders.

525. Marrinson, S. A. Territorial aspects of human
     distancing behavior, conceptualized within a
     social learning theory framework. Dissertation
     Abstracts International, 1978, 38-B, 3959.

526. Schmidt, J. R. Territorial invasion and aggression.
     Dissertation Abstracts International, 1976, 37-B,
     2578.

See also: citations 474, 1028, 1076, 1086, 1146.

e. Personal Space

527. Goffman, E. BEHAVIOR IN PUBLIC PLACES: NOTES ON THE
     SOCIAL ORGANIZATION OF GATHERINGS. Glencoe: Free
     Press, 1963, 248 pp.

Used the social and spatial etiquette of public places
to study the social organization of the public which
prescribes them. Used body movement and spatial behavior to

demonstrate the concept of involvement and the regulation of
its intensity and distribution by social norms.

528. Sommer, R. PERSONAL SPACE: THE BEHAVIORAL BASIS FOR
         DESIGN. Englewood Cliffs, New Jersey: Prentice-
         Hall, Inc., 1969. xi, 177 pp.

    A cornerstone of socio-spatial research. Brings
together many early studies by Sommer and his colleagues.
Analyzes socio-spatial behavior in specific settings and
provides descriptions of human spatial processes.

529. Aiello, J. R. A test of equilibrium theory: Visual
         interaction in relation to orientation, distance,
         and sex of interactants. Psychonomic Science,
         1972, 27, 335-336.

    An experimental study of visual interaction at varying
interpersonal distances and body orientations. Interactions
between sex and orientation and distance are reported:
females were found to be more visually reactive to others.

530. Aiello, J. R., & Thompson, D. E. Personal space,
         crowding and spatial behavior in a cultural
         context. In #11 (Altman et al., 4), pp. 107-178.

    A general review of theory and research on personal
space and crowding, with particular reference to the role of
culture and subculture.

531. Albert, S., & Dabbs, J. M., Jr. Physical distance and
         persuasion. Journal of Personality and Social
         Psychology, 1970, 15, 265-270.

    A field experiment analyzing the influence of physical
distance from a speaker on attitude change, inferred
expertise, and selective attention.

166   II. PSYCHOLOGICAL PROCESSES

532. Argyle, M., & Dean, J. Eye-contact, distance, and
     affiliation. Sociometry, 1965, 28, 289-304.

     A discussion of the interaction between eye-contact and
interpersonal distance. Outlines the concept of
compensation by one socio-spatial mechanism for changes in
the other to maintain an equilibrium level of contact
appropriate to the level of affiliation between persons.

533. Barefoot, J. C., Hoople, H., & McClay, D. Avoidance
     of an act which would violate personal space.
     Psychonomic Science. 1972, 28, 205-206.

     A field experiment probed people's willingness to
violate the personal space of others in order to use a
facility. Tested the hypothesis that spatial invasions are
aversive to the invader and are avoided.

534. Batchelor, J. P., & Goethals, G. R. Spatial
     arrangements in freely formed groups. Sociometry,
     1972, 35, 270-279.

     A study of social norms in terms of patterns of chairs
left by people asked to make group decisions compared to
those left by people making individual choices.

535. Bauer, E. A. Personal space: A study of blacks and
     whites. Sociometry, 1973, 36, 402-408.

     An experimental study of distances chosen as comfortable
between dyads of same-race-and-sex strangers.

536. Baum, A., Riess, M., & O'Hara, J. Architectural
     variants of reaction to spatial invasion,
     Environment and Behavior, 1974, 6, 91-100.

Used architectural barriers to probe the mediating
effects of the physical environment on interpersonal
spacing, and on reaction to invasion of personal space.

537.   Baxter, J. C.   Interpersonal spacing in natural
          settings. Sociometry, 1970, 33, 444–456.

The interpersonal spacing of varying ethnic, age, and
sex groupings in indoor and outdoor settings was
unobtrusively observed.

538.   Burgoon, J. K., & Jones, S. B.   Toward a theory of
          personal space expectations and their violations.
          Human Communication Research, 1976, 2, 131–146.

A review and theoretical analysis of problems of
personal space, presenting a model for conceptualizing
response to violations of personal space.

539.   Cheyne, J. A., & Effran, M. G.   The effect of spatial
          and interpersonal variables on the invasion of
          group-controlled territories. Sociometry, 1972,
          35, 477–489.

Reports two field studies of people's willingness to
invade the joint personal space of a dyad. Analyzes the
influences of the sex and composition of the pair being
invaded and their level of interaction and interpersonal
distance on people's avoidance of their shared space.

540.   Dabbs, J. M., Jr., & Stokes, N. A. III   Beauty is
          power: The use of space on the sidewalk.
          Sociometry, 1975, 38, 551–557.

A field experiment investigating the distance that
pedestrians deviate from their paths to circumvent

confederates varying in sex, number, and attractiveness.
These characteristics are interpreted in terms of their
representation of power relationships determining
appropriate distances.

541. Dosey, M. A., & Meisels, M. Personal space and self-
        protection. Journal of Personality and Social
        Psychology, 1969, 11, 93-97.

    Tested the hypothesis that people increase interpersonal
distance as a buffer against perceived threats in stressful
situations. Results supported the hypothesis in terms of
effects of experimentally manipulated stress, but found no
differences between groups high and low in anxiety.

542. Duke, M. P., & Nowicki, S. A new measure and social
        learning model for interpersonal distance.
        Journal of Experimental Research in Personality,
        1972, 6, 119-132.

    Reviews concepts of and methods for measuring spatial
behavior. Offers an alternative measure. Comfortable
Interperonal Distance (CID), based on a learning theoretic
model, and presents a series of experimental findings in
support of its validity.

543. Eberts, E. H., & Lepper, M. R. Individual consistency
        in the proxemic behavior of preschool children.
        Journal of Personality and Social Psychology,
        1975, 32, 841-849.

    Used spatial behavior of young children among peers and
with adults in structured and 'free play' settings to probe
the stability and cross-situation consistency of proxemic
behaviors.

544. Efran, M. G., & Cheyne, J. A. Affective concomitants
        of the invasion of shared space: Behavioral,

physiological, and verbal indications. Journal of
Personality and Social Psychology, 1974, 29,
219-226.

Compared non-verbal behavioral and cardiovascular
responses and self-reports of mood as indices of
individuals' response to violations of personal space,
observed through reactions to forced intrusions on
conversing pairs of persons.

545. Evans, G. W., & Howard, R. Personal space.
      Psychological Bulletin, 1973, 80, 334-344.

A review of the empirical, methodological and
theoretical literature on personal space, including
relationships to personality, pathology, and culture.

546. Fisher, J. D., & Byrne, D. Too close for comfort:
      Sex differences in response to invasions of
      personal space. Journal of Personality and Social
      Psychology, 1975, 32, 15-21.

A field study providing support for the hypothesis that
invasions of personal space from the side are more
threatening to women while frontal invasions are more
threatening to men. Observational evidence of the use of
personal belongings to shield against lateral or frontal
invasions was also analyzed.

547. Hayduk, L. Personal space: An evaluative and
      orienting overview. Psychological Bulletin, 1978,
      85, 117-134.

A literature review stressing methodological issues,
including measurement techniques. Contains an extensive
bibliography.

548. Hendrick, C., Giesen, M., & Coy, S. The social
        ecology of free seating arrangements in a small
        group interaction context. Sociometry, 1974, 37,
        262-274.

      A pair of field experiments explored the seating pattern
chosen around a group leader and the influence on those
patterns of a confederate who sat unusually close to or far
from the leader.

549. Jones, S. E., & Aiello, J. R. Proxemic behavior of
        black and white first-, third-, and fifth-grade
        children. Journal of Personality and Social
        Psychology, 1973, 25, 21-27.

      An observational analysis of interpersonal distance and
axial orientation. Studied sex and age differences, as well
as those between blacks and whites. Discusses the early
learning of proxemic patterns.

550. Knowles, E. S. Boundaries around social space:
        Dyadic responses to an invader. Environment and
        Behavior, 1972, 4, 437-447.

      A field experiment testing the strength of boundaries
around personal space shared by pairs of individuals. The
influences of sex and composition of the pair on their
willingness to avoid violation of their joint space were
probed.

551. Knowles, E. S. Boundaries around group interaction:
        The effect of groups size and member status on
        boundary permeability. Journal of Personality and
        Social Psychology, 1973, 26, 327-331.

      Extends the preceding to individuals interacting in
groups, and introduces perceived social status as a variable
affecting the extent to which boundaries are penetrated.

552. Knowles, E. S., & Bassett, R. L. Groups and crowds as social entities: Effects of activity, size, and member similarity on non-members. Journal of Personality and Social Psychology, 1976, 34, 837-845.

Well described by its title, this field experiment explored the amount of attention paid by passersby and changes in their expected behavior in relation to groups and crowds of varying characteristics. Reponses to the gatherings of people before, during and after passing them were analyzed.

553. Konecni, V. J., Libuser, L., Morton H., & Ebbesen, E. Effects of a violation of personal space on escape and helping responses. Journal of Experimental Social Psychology, 1975, 11, 288-299.

A series of four field experiments was designed to probe victims' behavior after invasion of their personal space while they were standing still, moving, or in a combination of both in situations that offered the victim the opportunity to aid the person who had invaded their personal space.

554. Mahoney, E. R. Compensatory reactions to spatial immediacy. Sociometry, 1974, 37, 423-431.

Used a field experiment to test the validity of spatial behaviors that in past experiments have been assumed to represent invasions of personal space.

555. Maines, D. R. Tactile relationships in the subway as affected by racial, sexual, and crowded seating situations. Environmental Psychology and Nonverbal Behavior, 1977, 2, 100-108.

An observational study on touching behavior on the part
of subway riders, focusing on the interactive roles of sex,
race, and crowding.

556. Mehrabian, A., & Diamond, S. G. Effects of furniture
     arrangement, props, and personality on social
     interaction. Journal of Personality and Social
     Psychology, 1971, 20, 18-30.

Used amount of conversation and demonstrations of
positive affect between strangers as measures of the
influence of interpersonal distance and orientation
resulting from furniture placement. The role of
conversation pieces was also explored.

557. Mehrabian, A., & Diamond, S. G. Seating arrangement
     and conversation. Sociometry, 1971, 34, 281-289.

Develops the concept of immediacy of spatial behavior
reflecting combinations of interpersonal distance, body
orientation, and non-verbal behavior. Describes an
experiment probing patterns of seating preference for
conversation as influenced by sex and personality variables.
Suggests a two-sided relationship between affiliation and
immediacy of spatial behavior.

558. Nesbitt, P. D., & Steven, G. Personal space and
     stimulus intensity at a southern California
     amusement park. Sociometry, 1974, 37, 105-110.

Probed the hypothesis that interpersonal spacing is used
to control the intensity of stimulation received from
others. Stimulation levels were operationalized as color
and scent in this field experiment.

559. Roos, P. Jurisdiction: An ecological concept. In #29
     (Proshansky et al., 1970), pp. 239-246.

Contrasts the concept of a temporary, frequently task-specific authority over or claim to space with theories of territoriality involving more lasting demarcation and defense of an area. Used a case study to demonstrate the dynamic aspects of spatial jurisdiction.

560. Russo, N. F. Eye-contact, interpersonal distance and the equilibrium theory. Journal of Personality and Social Psychology, 1975, 31, 497-502.

An experimental probe of the influences of sex, friendship, and interpersonal distance on the total amount and the mean duration of eye contact among elementary-school children. Interpersonal distance affected amount of eye contact, but not length of such contact. The reverse was found for friendships. Both variables were related to sex.

561. Sensenig, J., Reed, T., & Miller, J. Cooperation in the prisoner's dilemma as a function of interpersonal distance. Psychonomic Science, 1972, 26, 105-106.

Investigated the influence of subjects' seating pattern on levels of cooperation, levels of earnings during the game, and parity of outcome between players.

562. Smith, R. J., & Knowles, E. S. Affective and cognitive mediators of reactions to spatial invasions. Journal of Experimental Social Psychology, 1979, 15, 437-452.

A field study of response to "invasion" of personal space by a stranger, carried out at a street crosswalk. Results gave support to both arousal and cognitive-attributional explanations of responses in such a situation.

563. Sobel, R. B., & Lillith, N. Determinants of non-stationary personal space invasion. Journal of Social Psychology, 1975, 97, 39-45.

Used a Manhattan sidewalk as field setting to study patterns of collision avoidance by pedestrians whose paths were invaded. On only 60 percent of the trials did the pedestrians actually avoid a brush with the oncoming "invader," this happened more frequently in response to female than to male invaders.

564. Sommer, R. Leadership and group geography. Sociometry, 1961, 24, 99-110.

A classic observational field study of seating patterns chosen by small groups with and without leaders. Also discusses seating arrangements chosen by pairs of subjects at varying distances from each other.

565. Sommer, R. Further studies of small group ecology. Sociometry, 1965, 28, 337-348.

Reports a series of studies investigating seating patterns in public settings which encourage and settings which discourage social interaction. Outlines the influences of coacting vs. competitive purposes on preferred seating patterns.

566. Stratton, L. O., Tekippe, D. J., & Flick, G. L. Personal space and self-concept. Sociometry, 1973, 36, 424-429.

Compared comfortable distances of approach to a male confederate and a clothed dummy with placements of cutout figures for subjects rated high, medium and low on self-concept scales. Interactions of sex with self-concept in interpersonal approach were found.

567. Sundstrom, E., & Sundstrom, M. G. Personal space invasion: What happens when the invader asks

permission? Environmental Psychology and
Nonverbal Behavior, 1977, 2, 76-82.

Studied responses to an "invader" seating him-or herself
9 or 18 inches away from a subject on a campus bench in
terms of compensatory or avoidance reactions (interposing
barriers, looking away), and length of time until the
subject left.

568. Tennis, G. H., & Dabbs, J. M., Jr. Sex, setting, and
      personal space: First grade through college.
      Sociometry, 1975, 38, 385-394.

Experimentally probed the role of setting ("center
stage" vs. corner) on interpersonal distances chosen for
comfortable conversation by same-sex pairs at five age
levels. Interpersonal distance was higher in older as
compared to younger children, in males as compared to
females, and in corner compared to center stage settings.

569. Thayer, S., & Alban, L. A field experiment on the
      effect of political and cultural factors on the
      use of personal space. Journal of Social
      Psychology, 1972, 88, 267-272.

Probed the influence of political symbols (American flag
vs. peace button) on the interpersonal distance chosen by
residents of liberal and conservative neighborhoods when
asked for information.

570. Worthington, M. Personal space as a function of the
      stigma effect. Environment and Behavior, 1974, 6,
      289-294.

A field experiment concerning the influence of a
simulated physical disability on distances of interpersonal
approach by strangers in a public setting.

571. Edwards, H. F. Personal space invasion: The
      influence of confounding variables on arousal,
      productivity, and satisfaction. Dissertation
      Abstracts International, 1978, 38-B, 3466.

572. Kerr, M. K. Patterns of social group structure,
      interpersonal spacing and behavior in young
      children. Dissertation Abstracts International,
      1977, 37-B, 5437.

573. Saur, M. S. The effects of adaptation-level on
      responses to interaction distance. Dissertation
      Abstracts International, 1978, 38-B, 3906.

574. Vawter, J. M. Kindergarten children's expression of a
      personal space preference in the classroom and
      their spacing behavior in classrooms of varying
      densities. Dissertation Abstracts International,
      1977, 38-A.

See also: citations 504, 946, 1280, 1371.

6. BEHAVIORAL ECOLOGY

The ecological approach to the study of behavior focuses upon analyses of behavior in terms of situational context and the distribution of specific behaviors over different locales. Such work has been primarily identified with the school of behavioral ecology developed by Roger Barker and Herbert Wright at the University of Kansas, who, with the aid of a small army of devoted disciples, have been carrying on a vigorous program of research on relationships between "behavior settings" and human activities. While these behavioral ecologists have been inclined to define behavioral settings in institutional rather than physical terms, the approach nevertheless occupies an important place on the environment-behavior scene, especially the work that has dealt with community and institutional size, and the concept of under-versus over-manned settings.

The material included here is drawn primarily from the work of the Kansas group, but a few other works which are clearly ecological in tone, in terms of a primary focus on the environmental context of behavior, are included.

575. Barker, R. G. ECOLOGICAL PSYCHOLOGY: CONCEPTS AND METHODS FOR STUDYING THE ENVIRONMENT OF HUMAN BEHAVIOR. Stanford, Calif.: Stanford University Press, 1968. 242 pp.

A systematic presentation of the theory and methods of ecological psychology, illustrated through research conducted by the author and his collaborators at the Midwest Psychological Field Station.

576. Barker, R. G. & Associates. HABITATS, ENVIRONMENTS AND HUMAN BEHAVIOR: STUDIES IN ECOLOGICAL PSYCHOLOGY AND ECO-BEHAVIORAL SCIENCE. San Francisco: Jossey-Bass, 1978. xxxi, 327 pp.

177

An overview of the theory and research of Barker, his colleagues and disciples, introducing a distinction between "ecological psychology," which deals with individual behavior, and "eco-behavioral science," which involves the study of behavior settings and group behavior in relation to such settings.

577.   Barker, R. G., & Gump, P. V.   BIG SCHOOL, SMALL
          SCHOOL.   Stanford, Calif.:   Stanford University
          Press, 1964.   250 pp.

A report presented within the framework of behavior setting theory, delineating the relationship between the sizes of communities and of their high schools and the participation of adolescents in activities in and out of school.

578.   Barker, R. G., & Schoggen, P.   QUALITIES OF COMMUNITY
          LIFE.   San Francisco: Jossey-Bass, 1973, 562 pp.

A highly detailed presentation of the application of behavior-setting theory to the comparison of human behavior and activity in two small towns, one in the U.S.A., the other in England.   Provides an in-depth introduction to the raw data, as well as the methodological and conceptual features, of behavioral ecology.

579.   Barker, R. G., & Wright, H. F.   MIDWEST AND ITS
          CHILDREN.   Evanston, Illinois: Row Peterson,
          1955.   532 pp.

An early account of the application of the principles of behavioral ecology to the study of the lives and activities of children in a small rural Midwestern community.

580. Bechtel, R. ENCLOSING BEHAVIOR. Stroudsburg,
      Pennsylvania: Dowden, Hutchinson, and Ross, 1977.
      192 pp.

    An application of principles and techniques of
behavioral ecology to problems in the design of housing and
the built environment in general, starting from the premise
that architectural space "encloses behavior."

581. Bronfenbrenner, U. THE ECOLOGY OF HUMAN DEVELOPMENT:
      EXPERIMENTS BY NATURE AND DESIGN. Cambridge,
      Mass.: Harvard University Press, 1979. xv, 319
      pp.

    This volume presents a blueprint for an ecological
approach to the study of child behavior in context, i.e., in
the setting of home, family and community (i.e., at the
levels of meso-, exo- and macrosystems) in which development
takes place. Primary emphasis is on the role of
interpersonal relations and institutional forces.

582. Rogers-Warren, A., & Warren, S. F. (Eds.), ECOLOGICAL
      PERSPECTIVES IN BEHAVIOR ANALYSIS. Baltimore:
      University Park Press, 1976. 249 pp.

    Proceedings of a Conference designed to confront
representatives of the behavioral-ecology approach, such as
Gump and Willems, with proponents of behavior analysis.
Deals only incidentally with issues of the physical
environment in relation to behavior.

583. Sobal, J. Ecological psychology: an introduction and
      bibliography. Man-Environment Systems, 1976, 6,
      201-207.

584. Wicker, A. W. AN INTRODUCTION TO ECOLOGICAL
      PSYCHOLOGY. Monterey, Calif.: Brooks/Cole. 1979.
      xii, 228 pp. (paperback)

Covers theory and research on behavior settings,
behavior-environments congruence, under-and overmanning, and
organization size.

585. Baird, L. L. Big school, small school: A critical
      examination of the hypothesis. Journal of
      Educational Psychology, 1969, 60, 253-260.

Confirmed Barker and Gump's findings of the inverse
relationship between high-school size and amount of
participation in extra-curricular activities. Activities of
students in college were more closely related to size of the
college than to that of the previous high school.

586. Barker, R. G. Ecology and motivation. In M. R. Jones
      (Ed.), NEBRASKA SYMPOSIUM ON MOTIVATION (Vol. 8).
      Lincoln: University of Nebraska Press, 1960. Pp.
      1-48.

A rare instance of an attempt by a major representative
of behavioral ecology to deal with the problem of the
dynamics of behavior.

587. Barker, R. On the nature of the environment. Journal
      of Social Issues, 1963, 19(4), 17-38.

An examination of behavior-environment relations from
the perspective of behavioral ecology, emphasizing the
situational, environmental and behavior-setting
determination of behavior, as opposed to the role of person-
centered processes.

588. Barker, R. G. Explorations in ecological psychology.
      American Psychologist, 1965, 20, 1-14.

A brief statement of the main concepts of ecological
psychology and the premises of the theory of environment -
behavior relationships underlying this branch of psychology.

589.   Barker, R. G. The influence of frontier environments
          on behavior. In J. O. Steefen (Ed.), THE AMERICAN
          WEST, NEW PERSPECTIVES, NEW DIMENSIONS Norman,
          Oklahoma: University of Oklahoma Press, 1979.
          Pp. 61-93.

An application of the perspective of the behavioral
ecologist to an analysis of the conditions of living on the
American Western frontier.

590.   Bronfenbrenner, U. Toward an experimental ecology of
          human development. American Psychologist, 1977,
          32, 513-531.

A plea for a broader approach to the study of human
development, focusing on the changing interrelationships
between the organism and its environment, conceived in
system terms. Emphasis is on interpersonal and social
contexts of development.

591.   Ebbensen, E. B., Kjos, G. L., & Konecni, V. J.
          Spatial ecology: Its effects on the choice of
          friends and enemies. Journal of Experimental
          Social Psychology, 1976, 12, 505-518.

Investigated the relationships of distance between
residences, frequency of face-to-face contact and
probabilities of being named as a liked or disliked person.
Overall, friendship was related to frequency of contact
while dislike was more strongly influenced by actions that
lowered the experienced quality of life.

592.   Eddy, G. L., & Sinnett, E. R. Behavior setting
          utilization by emotionally disturbed college

students. Journal of Consulting Clinical
Psychology, 1973, 40, 210-216.

An application of the behavior-setting approach to
analyze the behavior patterns of an emotionally disturbed
group.

593. Forgas, J. P., & Brown, L. B. Environmental and
behavioral cues in the perception of social
encounters: An exploratory study. American
Journal of Psychology, 1977, 90, 635-644.

Studies the respective roles of interpersonal cues
provided by two persons shown interacting in dyads (via
slides), and of the environmental cues provided by the
setting as mediators of judgments of the social encounters
portrayed, with both the interacting dyads and the setting
shown in an intimate and a non-intimate version. The
results pointed to major interaction effects between the two
variables.

594. Gump, P. V. The behavior setting: A promising unit
for environmental designers. Landscape
Architecture. 1971, 61(2), 130-134.

A brief explanation of the behavior-setting concept and
of its application to the concerns of the designer.

595. Gump, P. V. Milieu, environment, and behavior. Design
and Environment, 1971, 2(4), 49ff.

A statement of the essentials of behavioral ecology and
its relevance for those in the design professions.

596. Gump, P. V. Operating environments in open and
traditional schools. School Review, 1974, 82,
575-593.

Concepts and methods relating to the open classroom, and discussion of its advantages and limitations compared to traditional classroom design.

597. Gump, P. V. Environmental psychology and the behavior setting. In #39 (EDRA-6), pp. 152-163.

Differentiates between descriptions of the environment based on inhabitant reaction and those based on the external world. Among the former are included phenomenological and objective (but subject-derived) modes of description, and those based on the use of reinforcement techniques. Among the latter are the meteorological, geographic and behavior-setting environments. A plea is made for the last-mentioned of these, and for the related concept of "synomorphs," representing the union of milieu and behavior program.

598. Gump, P. V., & Ross, R. The fit of milieu and programme in school environments. In H. McGurk (Ed.), ECOLOGICAL FACTORS IN HUMAN DEVELOPMENT, Amsterdam: North Holland, 1977. Pp. 77-89.

An analysis of an open class-room school in terms of the concept of synomorphy, i.e., congruence between the physical milieu and the standard behavior pattern occuring within it.

599. Ittelson, W. H., Rivlin, L. G., & Proshansky, H. M. The use of behavioral maps in environmental psychology. In #30 (Proshansky et al. 1976), pp. 340-351.

A descriptive study of the different forms of behavior observed on a psychiatric ward and their ecological distribution over different rooms and areas of the ward.

600. LeComte, W. F. Behavior settings as data-generating units for the environmental planner and architect. In #1472 (Lang), pp. 183-193.

A brief presentation of the behavior-setting approach
and its relevance for the environmental designer.

601. Ledbetter, C. B. Undermanning and architectural
        accessibility. In #38 (EDRA-5), Vol. 8, pp.
        281-288.

Presents data on occupancy of behavior settings in three
remote military stations in Alaska, which point to the role
of architectural accessibility of a setting as a possible
modulator of the effects of undermanning.

602. Moore, R., & Young, D. Childhood outdoors: Towards a
        social ecology of the landscape. In #10 (Altman
        and Wohlwill, 3), pp. 83-130.

A survey of work on children's behavior outdoors, in
play areas, neighborhood streets, etc., analyzed in
ecological terms, by focusing on locational and range
aspects of children's activities.

603. Petty, R. M. Experimental investigation of
        undermanning theory. In #38 (EDRA-5) Vol. 8, pp.
        259-269.

A review of several experimental studies.

604. Price, R., & Blashfield, R. Explorations in the
        taxonomy of behavior settings: Analysis of
        dimensions and classification of settings.
        American Journal of Community Psychology, 1975, 3,
        335-351.

An attempt to establish a classification scheme for
behavior settings, via factor- and cluster-analysis of a

particular set of behavior settings. The classifications
are predominantly in terms of the main groups occupying the
settings, e.g., children; adolescents; women.

605. Regnier, V. A. Matching older persons' cognition with
         their use of neighborhood areas. In #38 (EDRA-5,
         Vol. 11), pp. 19-40.

    Relates older persons' cognitive maps of urban
neighborhoods to their use of shopping and service
facilities in these areas.

606. Schoggen, P. & Barker, R. G. Ecological factors in
         development in an American and an English small
         town. In H. McGurk (Ed.), ECOLOGICAL FACTORS IN
         HUMAN DEVELOPMENT, Amsterdam: North-Holland, 1975.
         pp. 61-76.

    A comparison in terms of territorial range, behavioral
output, leadership responsibilities, etc. of the activities
of individuals ranging in age from infancy to old age living
in two small towns, one in England and one in the U.S.A.

607. Steinitz, C. Meaning and the congruence of urban form
         and activity. American Institute of Planners
         Journal, 1968, 34, 233-248.

    Formal characteristics of locations in downtown Boston
were related to dominant activities occurring there to
determine the extent of form-activity congruence. These
congruence data were related in turn to meaningfulness
values and imaging data.

608. Wicker, A. W. Undermanning, performances, and
         students' subjective experiences in behavior
         settings of large and small high schools. Journal
         of Personality and Social Psychology, 1968, 10,
         255-261.

Comparisons between and within two high schools
differing in size pointed to the role of undermanning as a
mediator of students' sense of satisfaction, self-confidence
and challenge provided by their extra-curricular activities.

609. Wicker, A. W. Undermanning theory and research:
      Implications for the study of psychological and
      behavioral effects of excess populations.
      Representative Research in Social Psychology,
      1973, 41, 185-206.

     A general presentation of the concept of undermanned
behavior settings and related aspects of behavioral ecology.

610. Wicker, A. W., & Kirmeyer, S. From church to
      laboratory to national park: A program of
      research on excess and insufficient populations in
      behavior settings. In #54 (Wapner et al.) pp.
      157-185; also in #13 (Stokols), pp. 69-99.

     An application of the concepts of over- and undermanned
behavior settings to the analysis of diverse phenomena in
both laboratory and field settings. Includes data on church
members' participation in the activities of their
congregation, and on the behavior of the staff of an
overcrowded National Park in coping with the demands made on
them.

611. Willems, E. P. Behavioral ecology and experimental
      analysis: Courtship is not enough. In J. R.
      Nesselroade and H. W. Reese, (Eds.), LIFE-SPAN
      DEVELOPMENTAL PSYCHOLOGY: METHODOLOGICAL ISSUES.
      New York: Academic Press, 1973. Pp. 197-218.

     A consideration of behavior, and of the effects of
attempts to modify behavior, from the perspective of a
dynamic behavioral ecology, emphasizing the
interrelationship among behavioral forces and the

unpredicted, unwanted side-effects or longer-range impacts
of such intervention.

612. Willems, E. P. Behavioral ecology. In #13 (Stokols),
         pp. 39-68.

     Outlines the major defining attributes of behavioral
ecology, with emphasis on the naturalistic character of the
approach, the site-specific and place-dependent nature of
behavior, the molar nature of the phenomena studied and the
transdisciplinary feature of behavioral ecology.

613. Wright, H. F. Psychological development in the
         Midwest. Child Development, 1956, 27, 265-286.

     An ecological study of the occupancy distribution of
different family and community behavior settings in a small
rural town, as a function of the individual's age level
(from infancy to old age), mode of participation, and degree
of penetration into different settings at each age level.

614. Adelberg, B. Activity ranges of children in urban and
         exurban communities. Dissertation Abstracts
         International, 1978, 38-B, 3462.

615. Bechtel, R. B. Footsteps as a measure of human
         preference. Dissertation Abstracts, 1968, 28-B,
         3458-3459.

616. Deeble, C. T. Cognitive elements of the child's life
         situation in communities differing in size.
         Dissertation Abstracts, 1964, 24, 582-5527.

617. Donmez, A., Behavior setting comparison of a Turkish
         town and an American town. Dissertation Abstracts
         International, 1979, 39-B, 4101.

618. Perkins, D. V. Manning theory and the person-
      environment model of behavior: An empirical
      evaluation. Dissertation Abstracts International,
      1979, 39-B, 3534-3535.

See also: citations 956, 1036, 1038, 1040, 1041, 1255.

PART III

APPLIED AREAS AND SPECIAL ISSUES

As noted in the introduction, the environment-and-behavior field owes whatever claims it may have to autonomous status, independent of the parent fields of psychology, sociology, architecture and design, etc., to its interdisciplinary, as well as problem-focused character. Much of the work in the field has been inspired, not so much from an interest in behavioral processes, or in phenomena of environment-behavior relations, as from practical problems arising in specific settings, or from specific issues, demanding the attention of the behavioral scientists. (See Altman, item #56, for a persuasive discussion of the disparity between the social scientist's emphasis on process, as opposed to the practitioner's orientation towards problem solving, and towards particular settings.)

It is this aspect of the field that provides the material for Part III of our volume, which is arranged according to a combination of settings (natural environments, urban settings, institutions) and practical concerns (environmental problems, transportation, demographic issues).

7. THE NATURAL ENVIRONMENT AND BEHAVIOR

The writings of philosophers, poets and naturalists, the works of painters of all periods and styles, the zealous efforts of conservationists and environmentalists all bear witness to the high amount of affect and interest that we invest in our natural environment. Yet this topic has played little role in the development of environmental psychology as it is typically conceived, as a perusal of major reviews (e.g., item #6), and anthologies (e.g., item #30) makes apparent. It is as though psychologists, preoccupied with concepts of proxemics and socio-spatial processes, with problems of environmental stress, and with cognitive representation of the environment, have not known quite what to do with this somewhat elusive side of our environmental experience. This has not, however, inhibited those in diverse applied fields, including landscape architecture, forestry, natural recreation, and natural resource conservation, from devoting themselves to the study of behavioral and social-science aspects of problems relating to the use, conservation, and management of the natural environment, aided by a scattering of psychologists, sociologists and even economists who have harbored compatible interests.

There is, thus, a fairly extensive literature in this field, most of which can be subsumed under two headings: the perception and assessment of landscape and scenery, and attitude and behavior related to the use of natural settings for recreational purposes. In addition, there is an all too limited amount of material on the function of natural elements in our residential and urban areas (e.g., in the form of plants, gardens, playgrounds, and parks), and a quite different literature, developed largely through the efforts of a group of geographers, dealing with perception and response to natural hazards.

a. Gardens, Parks, Playgrounds and General Response to
Greenery and Vegetation

619. CHILDREN, NATURE, AND THE URBAN ENVIRONMENT:
          Proceedings of a Symposium-Fair. (USDA Forest
          Service General Technical Report NE-30).
          Washington, D. C.: U.S. Government Printing
          Office, 1977. x, 261 pp.

     A set of 35 predominantly brief papers, partly essays,
partly reports of empirical studies, concerning diverse
aspects of children's response to the world of nature and of
cities.

620. Cook, J. A. Gardens on housing estates: A survey of
          user attitudes and behavior on seven layouts.
          Town Planning Review, 1968, 39, 217-234.

     Reports a survey of over 600 residents of housing
developments in England (including New Towns), with respect
to satisfaction with and use of gardens. The role of garden
size and privacy is considered.

621. Driver, B. L., & Greene, P. Man's nature: Innate
          determinants of response to natural environments.
          In #619 (U. S. Forest Service, 1977), pp. 63-70.

     A speculative discussion of the biological and
evolutionary roots of man's response to stimuli of the
natural environment.

622. Gold, S. M. Nonuse of neighborhood parks. Journal of
          the American Institute of Planners, 1972, 38,
          369-378.

     A general discussion of the problem of underuse of
neighborhood parks by urban residents. Evidence of such

nonuse is reviewed, and the role of diverse behavioral, environmental and institutional factors as contributors to this problem is considered.

623.   Gold, S. M.   Neighborhood parks: The nonuse
           phenomenon.   Evaluation Quarterly, 1977, 2,
           319-327.

Reasons for the failure of people to make use of neighborhood parks are reviewed, and suggestions offered for remedying this problem.

624.   Gold, S. M.   Social and economic benefits of trees in
           cities.   Journal of Forestry, 1977, 75, 84-87.

Users of six neighborhood parks in Sacramento were interviewed to determine park features associated with user satisfaction.   For four of the six parks, the presence of trees was found to constitute the most important feature.

625.   Holcomb, B.   The perception of natural vs. built
           environments by young children.   In #619 (U. S.
           Forest Service, 1977), pp. 33-36.

A brief discussion and report of an exploratory study of four year old children's response to the natural as opposed to the built environment.   The distinction is regarded as irrelevant to the young child, who derives satisfaction from properties (opportunity for exploration, manipulation, collection of objects) encountered in both types of settings.

626.   Kaplan, R.   Some psychological benefits of gardening.
           Environment and Behavior, 1973, 5, 145-162.

An exploratory study of sources of satisfaction from gardening activity on the part of residents of a large

university community, including both home gardening and use
of a community garden. Relationships between gardening and
environmental attitudes, as well as role variables, are
reported.

627. Kaplan, R. The green experience. In S. Kaplan & R.
        Kaplan (Eds.), HUMANSCAPE: ENVIRONMENTS FOR
        PEOPLE. North Scituate, Mass.: Duxbury Press,
        1978. Pp. 186-193.

     A general, non-technical discussion of people's response
to greenery in the environment.

628. Lewis C. A. People-plant interaction: A new
        horticultural perspective. American
        Horticulturalist, 1973, 52(1), 18-25.

     A discussion, profusely illustrated via color
photographs, of the personal involvement on the part of
residents of low-income urban neighborhoods and housing
projects in the cultivation and care of flowers and gardens,
and of the beneficial effect of such involvement on the
sense of community thus developed.

629. Lewis, C. A. People/plant proxemics: A concept for
        humane design. In #40 (EDRA 7), pp. 102-107.

     A general discussion of people's response to plants and
vegetation, contrasting perceptual and behavioral aspects.
Among the latter gardening is singled out as an activity of
particular importance, and results from a program of
gardening in low income housing are summarized.
Implications for designers are discussed.

630. Marans, R. W. Outdoor recreation behavior in
        residential environments. In #14 (Wohlwill &
        Carson), pp. 217-232.

A study of the determinants of participation in outdoor recreation on the part of residents of planned new towns and other suburban developments, comparing the role of socioeconomic, motivational (e.g., affiliative, exploratory and status needs) and environmental (e.g., distance, location, facilities) variables.

631. Thayer, R. L., Jr., & Atwood, B. G. Plants, complexity, and pleasure in urban and suburban environments. Environmental Psychology and Nonverbal Behavior, 1978, 3, 67-75.

Ratings of slides of various environmental settings, with and without vegetation, indicated that the presence of vegetation consistently increases experienced pleasure, but does not affect their perceived complexity.

632. Bonnett, P. A. Nature in the urban landscape: A review and bibliography. Unpublished Master's thesis. University of Alberta, Edmonton, Alberta (Canada), 1971.

633. Gallagher, T. J. Visual preference for alternative natural landscapes. Dissertation Abstracts International, 1977, 38-A, 1702.

See also: citations 78, 1275.

b. Perception and Assessment of Landscape and Scenery
(For more general material on perceptual and evaluative responses to the environment, see Chapters 3-a and 3-c.)

634. Arthur, L. M., & Boster, R. S. Measuring scenic beauty: A selected annotated bibliography. Ft.

Collins, Colo.: Rocky Mountain Forest and Range
Experiment Station, 1976. (U.S. Forest Service
General Technical Reports, RM-25), 34 pp.

635. Cerny, J. W. Landscape amenity assessment
         bibliography. Monticello, Ill.: Council of
         Planning Libraries, 1972. (Exchange
         Bibliographies No. 287.)

Contains 45 references (unannotated).

636. Daniel, T. C., Zube, E. H., & Driver, B. L. (Eds.)
         ASSESSING AMENITY RESOURCE VALUES. Ft. Collins,
         Colo.: Rocky Mountain Forest and Range Experiment
         Station, U. S. Dept. of Agriculture. (General
         Technical Report RM-68), 1979. 70 pp.

A set of papers on diverse aspects of landscape
aesthetics and other perceptual and behavioral aspects of
natural resources, including wildlife.

637. Dearden, P. Landscape aesthetics: An annotated
         bibliography. Monticello, Ill.: Council of
         Planning Librarians, 1977. (Exchange
         Bibliographies #1220) 19 pp.

Contains diverse materials relating to visual quality of
the landscape and its assessment. Less than one-half of the
items are annotated.

638. Zube, E. H., Brush, R. O., & Fabos, J. G. (Eds.)
         LANDSCAPE ASSESSMENT: VALUES, PERCEPTIONS AND
         RESOURCES. Stroudsburg, PA.: Dowden, Hutchinson
         and Ross, 1976. xii, 367 pp.

A collection of theoretical, methodological and empirical papers from diverse perspectives, including history, economics, landscape architecture, natural resource management, and psychology.

639. Allen, N. J. R. Environmental perception of niveous landscapes. Proceedings of the Eastern Snow Conference, 1975, 20, 24-35.

Used photographs of snow scenes varying in terms of type of terrain and vegetation and amount of use or crowding suggested by ski or snowmobile tracks. Preferences for such photos were compared for snowmobilers and skiers.

640. Arthur, L. M. Predicting scenic beauty of forest environments: Some empirical tests. Forest Science, 1977, 23, 151-160.

Comparison of judgments of slides of forest stands among landscape architects, students, and members of the general public revealed major differences in means, but high agreement in relative ordering of different areas. "Timber cruise" variables (e.g., tree density, percent crown cover, amount and distribution of slash) proved the best predictors of judged beauty, but other physical features and attributes of the photographic design were also important.

641. Arthur, L., Daniel, T. C., & Boster, R. S. Scenic assessment: An overview. Landscape Planning, 1977, 4, 109-129.

A survey of three types of techniques for evaluating scenic beauty of natural resources; descriptive inventories, public evaluations and economic analyses. Focus is on methodological adequacy and on utility for scenic resource management.

642.   Brush, R. O. & Shafer, E. L. Application of a
       landscape preference model to land management.
       In #638 (Zube et al.), pp. 168–182.

       By means of a photographic overlay technique, effects of
specified changes in landscape features on preference are
determined, through the application of a general landscape-
preference model.

643.   Buhyoff, G. J., & Leuschner, W. A. Estimating
       psychological disutility from damaged forest
       stands. Forest Science, 1978, 24, 424–432.

       Preference judgments for forest scenes that had been
subjected to varying levels of damage from insects were
studied via a paired-comparison procedure. Preference
functions dropped sharply up to a damage level of 10% of the
total area, with only a slight drop beyond that point.

644.   Cherem, G. J., & Traweek, G. J. Visitor employed
       photography: A tool for interpretive planning on
       river environments. In #684 (U.S. Forest Service,
       1977), pp. 236–244.

       Presents a methodology for studying affective responses
to landscapes by asking outdoor recreationists to photograph
whatever they considered appealing (for one group) or
unappealing (for another group). Preliminary data are
presented from the application of this method to a four-mile
section of the Huron River in Michigan.

645   Clamp, P. Evaluating English landscapes -- some
       recent developments. Environment and Planning-A,
       1976, 8, 79–92.

       Describes an approach to the evaluation of the
attractiveness of landscapes, and of the impact on perceived
attractiveness of a road built through the field of view,

using computer-graphic type simulations overlaid on
photographs.

646.   Cook, W. L., Jr.   An evaluation of the aesthetic
          quality of forest trees.   Journal of Leisure
          Research, 1972, 4, 293-302.

       A study of visitors' judgments of attractiveness of
different hardwood species of trees in three natural
recreation areas in the Eastern U.S.   Relationships to
timber quality, and to aesthetic factors such as balance,
straightness and complexity (branching) are reported.

647.   Craik, K. H.   Appraising the objectivity of landscape
          dimensions.   In #682 (Krutilla), pp. 292-346.

       A set of Landscape Rating Scales, based on the degree of
presence of specified characteristics or features, is
developed and tested for reliability; relationships to
judgments of aesthetic appeal are obtained, leading to the
development of a Landscape Adjective Check List.

648.   Craik, K. H.   Psychological factors in landscape
          appraisal.   Environment and Behavior, 1972, 4,
          255-342.

       A review of research on landscape assessment methods.

649.   Craik, K. H.   Individual variations in landscape
          description.   In #638 (Zube et al.), pp. 130-150.

       Based on factor analysis of ratings of a suburban
landscape, a typology of response to landscape based on
combinations of high vs. low placement on each of four
factors is derived.

650. Daniel, T. C., Anderson, L. M., Schroder, H. W., &
      Wheeler, L. III. Mapping the scenic beauty of
      forest landscapes. Leisure Sciences, 1977, 1,
      35-52.

Describes a psychophysically derived method, "Scenic
Beauty Estimation," for assessing the visual quality of
relatively homogeneous landscapes, such as stands of forest,
and illustrates its application to a Ponderosa pine forest.

651. Daniel, T. C., & Boster, R. S. Measuring landscape
      aesthetics: The scenic beauty estimation method.
      Fort Collins: Rocky Mountain Forest and Range
      Experiment Station, 1976. (USDA Forest Service
      Research Paper RM-167.)

Describes in detail the development, validation and
testing, and application of a method for estimating scenic
beauty of landscapes. The method adjusts ratings for
systematic biases of an individual in mean level of ratings,
and uses a stimulus-sampling approach.

652. Dunn, M. C. Landscape with photographs: Testing the
      preference approach to landscape evaluation.
      Journal of Environmental Management, 1976, 4,
      15-26.

A discussion of problems of landscape assessment,
including results of a study of preferences for
photographically portrayed rural landscapes, and reasons for
them. Preferences frequently reflected associations of
particular sites to crowding, or activities, rather than
explicit visual qualities.

653. Fabos, J. G An analysis of environmental quality
      ranking systems. In #683 (Larson), pp. 40-55.

Examines systems in common use for assessment of
environmental quality, under three categories: resource
evaluation for policy planning; resource evaluation for
planning at a regional, state and subregional level; and
landscape evaluation for a specific purpose, such as camping
or boating. Systems based on rankings of preference for
specified landscape characteristics are included under the
first two.

654 Fabos, J. G. et al. Model for landscape resource
        assessment. Amherst, Mass.: University of
        Massachusetts, College of Food and Natural
        Resources, 1973. x, 141 pp. (Research Bulletin
        #602.)

   Describes a comprehensive model for the assessment of
landscape resource quality, including water quality and
supply, wildlife and agricultural productivity and visual
land-use complexity and compatibility.

655. Fines, K. D. Landscape evaluation: A research
        project in East Sussex. Regional Studies, 1968,
        2, 41-55.

   Develops a scale of landscape values for application to
both natural landscapes and townscapes.

656. Gauger, S. E., & Wyckoff, J. B. Aesthetic preferences
        for water resource projects: An application of Q
        methodology. Water Resources Bulletin, 1973, 9,
        522-528.

   A Q-sort of photographs of water development projects
revealed two factors, one involving man- vs. nature-
dominance, the other a combination of incomplete and
polluted appearance. Differences between photographers and
town assessors were found.

657.   Greenbie, B. B.   Problems of scale and context in
       assessing a generalized landscape for particular
       persons.   In #638 (Zube et al.), pp. 65-91.

    A discussion of the role of scale and context in
modulating response to landscape.   Includes an attempt to
apply sentics (the structural analysis of emotional
responses to a stimulus) to the realm of landscape.

658.   Hamill, L.   Statistical tests of Leopold's system for
       quantifying aesthetic factors among rivers.   Water
       Resources Research, 1974, 10, 395-401.

    Determined correlations between evaluative ratings of
riverine landscapes and the uniqueness ratios derived by
Leopold (cf. items 664, 665).   Correlations were poor for
physical and water quality features, better for human use
and scenic factors.   A critique of Leopold's methodology is
presented.

659.   Hamill, L.   Analysis of Leopold's quantitative
       comparisons of landscape aesthetics.   Journal of
       Leisure Research, 1975, 7, 16-28.

    A critique of Luna Leopold's approach (see items #664,
665) to evaluating the aesthetic value of a natural
resource, focusing on the calculation of a "uniqueness
ratio" and related problems.

660.   Hendrix, W. G., & Fabos, J. G.   Visual land use
       compatibility as a significant contributor to
       visual resource quality.   International Journal of
       Environmental Studies.   1975, 8, 1-8.

    Examined the perceived compatibility of adjacent land-
uses of different kinds, by obtaining compatibility ratings
from pairings of photographs representing diverse land-uses,
from undeveloped natural scenes to examples of industrial,
transportation and residential development.

661.   Jacobs, P. The landscape image: Current approaches
       to the visual analysis of the landscape. <u>Town
       Planning Review</u>, 1975, <u>46</u>, 127-150.

A discussion of issues in the systematic analysis and
assessment of landscapes, including problems in the analyses
of independent and dependent variables, selection of
prototype models for calibration, and issues of evaluation
and application of particular approaches. Seven case
studies of different models of assessment are reviewed.

662.   Kaplan, R. Down by the riverside: Informational
       factors in waterscape preference. In #684 (U.S.
       Forest Service, 1977), pp. 285-289.

Applies the Kaplan informational approach to the study
of environmental preference to riverine landscapes.

663.   Krieger, M. H. What's wrong with plastic trees?
       <u>Science</u>, 1973, <u>179</u>, 446-455.

A discussion of human conceptions and values regarding
the world of nature, with specific reference to (a)
assumptions implicit in attempts to "improve upon" the
appearance of the natural environment through "cosmetic"
intervention, and (b) the history of attitudes towards and
support for natural resource conservation.

664.   Leopold, L. B. Landscape esthetics. In A. Meyer
       (Ed.), ENCOUNTERING THE ENVIRONMENT. New York:
       Van Nostrand, 1971. Pp. 29-46.

An attempt to derive a method to assess the aesthetic
value of a natural area such as a section of a riverbed,
through the determination of "uniqueness ratios" for
specified dimensions of the landscape, including both

objective, topographically defined and perceptual
dimensions.

665.   Leopold, L. B., & Marchand, M. O.   On the quantitative
       inventory of the riverscape. Water Resources
       Research, 1968, 4, 709-717.

    Develops a method of assessing the aesthetic value of a
particular natural setting by means of a "uniqueness ratio,"
based on the degree of communality of the rating of that
setting on various physical and evaluative scales with
corresponding ratings of other similar settings. The use of
the method is described with reference to a set of 24
valleys in the vicinity of Berkeley, California.

666.   Lime, D. W., & Stankey, G. H.   Carrying capacity:
       Maintaining outdoor recreation quality. In #683
       (Larson) pp. 174-184.

    A discussion of the concept of recreational carrying
capacity, in terms of the dual impact of recreational use on
the quality of the resource, and the experience of the user.
Alternative strategies for management to optimize both
resource quality and user satisfaction are presented.

667.   Litton, R. B. Jr.   Aesthetic dimensions of the
       landscape. In #682 (Krutilla), pp. 262-291.

    A descriptive analysis of attributes and features of
landscapes relevant to aesthetic response.

668.   Litton, R. B. Jr.   Visual vulnerability of forest
       landscapes. Journal of Forestry, 1974, 72,
       392-397.

    A discussion of the vulnerability, from an aesthetic
standpoint, of forests to manmade alterations in the form of

roads, timber harvest, etc. Argues for sensitivity on the
part of foresters to specific features affecting visual
appearance.

669. Peterson, G. L., & Neumann, E. S. Modeling and
     predicting human response to the visual recreation
     environment. Journal of Leisure Research, 1969,
     1, 219-237.

     Presents an information-processing model of preference
for the visual environment, and applies it to evaluative
judgments of photographically presented beach areas. A
study of users of an urban beach on Lake Michigan disclosed
two sharply differentiated groups, one preferring a purely
natural and uncrowded setting, and the other the opposite.

670. Shafer, E. L. Jr. Perception of natural environments.
     Environment and Behavior, 1969, 1, 71-82.

     A discussion of perception and preference for natural
environments used in outdoor recreation, based in part on
the author"s study (with R. C. Thompson) of campers' choices
of campgrounds (see #742).

671. Shafer, E. L., Hamilton, J. F. Jr. & Schmidt, E. A.
     Natural landscape preferences: A predictive
     model. Journal of Leisure Research, 1969, 1,
     1-19.

     Developed and tested a model to predict preferences for
photographically portrayed natural landscapes. A set of six
variables, involving immediate, intermediate and distant
vegetation (perimeter and area) and area of water and of
distant nonvegetation, accounted for two-thirds of the
variance.

672. Shafer, E. L., & Richards, T. A. A comparison of
     viewer reactions to outside scenes and photographs

of those scenes. U.S. Forest Service Research
Paper NE-302, 1974. (Northeast Forest Experiment
Station, Upper Darby, Pa.) 26 pp.

An attempt to validate the use of photographic
simulation of outdoor environments, by comparing ratings
given to eight scenes, presented via color slides, color
photos, as well as viewed in the field. Results consist
mainly of profiles of the mean ratings of each scene on 27
bi-polar scales under the three conditions, and suggest
overall good agreement among them. Also published in
abbreviated form in #47 (Canter and Lee), pp. 71-79.

673. Shafer, E. L., & Tooby, M. Landscape preferences: An
      international replication. Journal of Leisure
      Research, 1973, 5, 60-65.

Demonstrated the generalizability of Shafer's
mathematical model for landscape preferences (see item 671)
to a Scottish population. The validity of data based on
black-and-white photographs of landscapes is reinforced by
these results.

674. Sonnenfeld, J. Environmental perception and
      adaptation level in the arctic. In D. Lowenthal
      (Ed.), Environmental perception and behavior.
      Chicago: Department of Geography, University of
      Chicago, 1967 (Research Paper #109). Pp. 42-57.

A comparison of preference choices made between pairs of
slides varying in terms of specified features of the
landscape (e.g., presence of vegetation and water,
ruggedness of terrain, etc.) on the part of groups of
differing amounts of experience in an Arctic environment,
including Eskimos living in differing locales as well as
white natives working in the Arctic, and students in the
U.S.

675. Wenger, W. D., & Videbeck, R. Eye pupillary
      measurement of aesthetic responses to forest
      scenes. Journal of Leisure Research, 1969, 1,
      149-161.

      An attempt to study the individual's affective reactions
to forest scenes through measurement of pupillary dilation.
Differences between groups (e.g., campers vs. non-campers)
are reported.

676. Wohlwill, J. F., & Harris, G. Response to congruity
      or contrast for man-made structures in natural
      settings. Leisure Sciences, 1980, 3, 349-366.

      Slides of urban- and state-park settings were scaled
with respect to the congruity or contrast between the man-
made and natural elements of each. These measures proved to
be highly predictive of mean evaluative ratings given to
each slide by a group of undergraduate subjects, but failed
to correlate with the amount of time subjects spent
voluntarily exploring the slide.

677. Zube, E. H. Evaluating the visual and cultural
      landscape. Journal of Soil and Water
      Conservation, 1970, 25, 137-141.

      Reviews a study focused on water resources in the
Northeast, involving evaluation and inventorying of
landscape features, along with ratings of scenic value of
state parks, forests and historic sites.

678. Zube, E. H. Cross-disciplinary and intermode
      agreement on the description and evaluation of
      landscape resources. Environment and Behavior,
      1974, 6, 69-90.

      Ratings of a set of landscapes presented via aerial
photographs on diverse semantic-differential scales by a

group of resource managers and a group of environmental
designers agreed closely in terms of both mean values and
correlations. Good agreement between ratings obtained in
the field and from aerial photos is also reported.

679.   Zube, E. H., Pitt, D. G., & Anderson, T. W.
         Perception and prediction of scenic resource
         values of the Northeast. In #638 (Zube et al.),
         pp. 151-167.

   Describes an extensive investigation of perceived scenic
quality of scenes in the Connecticut River Valley, including
comparisons among different professional and nonprofessional
groups, between field and slide methodologies, and between
data based on evaluative (quasi-objective) and preference
(subjective) sets. Data on the contribution of various
topographical dimensions to the prediction of scenic quality
are presented.

680.   Hammitt, W. E. Visual and user preference for a bog
         environment. Dissertation Abstracts
         International, 1978, 39-A, 3833-3834.

See also: citations 220, 234, 841.

c. Outdoor Recreation Behavior in Natural Settings

681.   Hendee, J. C., & Schoenfeld, C. (Eds.) HUMAN
         DIMENSIONS IN WILDLIFE PROGRAMS. Rockville, MD:
         Mercury Press, 1973. 193 pp.

   A set of 18 papers on perception of and response to
wildlife, and attitudinal and behavioral issues in wildlife

management. Includes several papers on hunter satisfaction
and values, and anti-hunting attitudes.

682. Krutilla, J. V. (Ed.) NATURAL ENVIRONMENTS: STUDIES
         IN THEORETICAL AND APPLIED ANALYSIS. Baltimore:
         Johns Hopkins Press, 1972. 352 pp.

   A set of papers dealing with the perception, value, use
and management of natural environments, from the differing
perspectives of economics, ecology, natural resource
management, landscape architecture, and psychology.

683. Larson, E. V. H. (Ed.) THE FOREST RECREATION
         SYMPOSIUM. Proceedings of a Recreation Symposium,
         University of Syracuse, 1971. U. S. Forest
         Service, Northeastern Forest Experiment Station,
         Upper Darby, Pa., 1971. 211 pp.

   Includes a section, "Characterizing the recreation user"
(pp. 105-149), containing chapters on campers, hikers,
members of conservation organizations, hunters and
fishermen, skiers, and snowmobilers.

684. U. S. Forest Service, North Central Forest Experiment
         Station. PROCEEDINGS, SYMPOSIUM: RIVER
         RECREATION MANAGEMENT AND RESEARCH. St. Paul,
         Minn.: North Central Forest Experiment Station,
         1977. 455 pp.

   Includes diverse material of a behavioral science
nature, relating to motivational factors and sources of
satisfaction in river recreation, assessment and evaluation
of riverine landscape quality, response to crowding, user
attitudes towards management and methodology of river
recreation research. (See items #644, 662, 708.)

685. Barcus, C. G., & Bergeson, R. G. Survival training
     and mental health: A review. Therapeutic
     Recreation Journal, 1972, 6, 3-7.

     A brief review of selected studies of the effects of
Outward-Bound survival training programs on self-concepts,
academic achievement, and other aspects of behavior of
adolescents and college-age youth.

686. Becker, R. H. Social carrying capacity and user
     satisfaction: An experiential function. Leisure
     Sciences, 1978, 1, 241-258.

     A study of user characteristics and experiential and
attitudinal correlates related to preference for sites
within a state forest characterized by different levels of
user density.

687. Bryan, H. Leisure value systems and recreational
     specialization: The case of trout fishermen.
     Journal of Leisure Research, 1977, 9, 174-187.

     Demonstrates an increase in specialization (in terms of
equipment used, level of skill, and preference for setting)
in trout fishermen as a function of continued participation
in this sport, and a correlated shift from an emphasis on
consumption to one on preservation and environmental
setting.

688. Bultena, S. L., & Taves, M. J. Changing wilderness
     images and forestry policy. Journal of Forestry,
     1961, 59, 167-171.

     A survey of campers and canoers in the Quetico-Superior
wilderness, demonstrating the co-existence of conflicting
values and desires, for a "pure" wilderness experience on
the one hand and for facilities and amenities on the other.
The possible basis for this conflict and of the manner in
which it is dealt with by the user, is discussed.

689. Burch, W. R. Jr. Two concepts for guiding recreation management decisions. Journal of Forestry, 1964, 62, 707-712.

A discussion of individual- and social-behavioral aspects of natural recreation of importance for the recreation manager, with particular emphasis on collective (i.e., group-based) aspects of recreational behavior, and on the user's definition of the resource system.

690. Burch, W. R. Jr. The play world of camping: Research into the social meaning of outdoor recreation. American Journal of Sociology, 1965, 70, 604-612.

An examination of campers and their behavior from the perspective of the sociology of play, differentiating between different modes of play (symbolic, expressive, subsistence, etc.), and relating each to sex roles.

691. Burch, W. R. Jr. Wilderness -- the life cycle and forest recreational choice. Journal of Forestry, 1966, 64, 606-610.

Presents data relating the accessibility or remoteness of campgrounds to the stage of the life-cycle of the users of those campgrounds. Implications for the problem of devising outdoor recreation facilities for the elderly are considered.

692. Burgess, R. L., & Field, D. R. Methodological perspectives for the study of outdoor recreation. Journal of Leisure Research, 1972, 4, 63-72.

Discusses six perspectives for outdoor-recreation research: (a) social-aggregate (i.e., sociological, including role of demographic variables, activity patterns,

etc.), (b) social-psychological (attitudes, values) (c)
social-organizational (role of family, organized groups and
movements), (d) activity syndromes (typologies of users),
(e) community and regional analysis, and (f) social-
ecological (interrelations of elements of a complex system).

693.   Catton, W. Motivation of wilderness users. Pulp and
       Paper Magazine of Canada, December 19, 1969, 70,
       121-126.

A general discussion of the differing motivations
underlying the seeking out of wilderness experiences,
including consideration of the role of stress-seeking,
escape from tensions of everyday living, search for privacy,
and close interaction with companions.

694.   Cheek, N. H. Jr.   Variations in patterns of leisure
       behavior: An analysis of sociological aggregates.
       In #1447 (Burch et al.), pp. 29-43.

A report on demographic correlates of park-visiting
behavior.

695.   Cheek, N. H., Jr., & Field, D. R.   Aquatic resources
       and recreation behavior. Leisure Sciences, 1977,
       1, 67-83.

A survey conducted in Washington state, aimed at
inventorying the different kinds of recreational activities
engaged in by the respondents, in aquatic-resource areas.
Results demonstrate the extent to which such areas are used
for both aquatic and non-aquatic forms of recreation.
Implications for natural-resource and recreation management
are discussed.

696.   Cicchetti, C. J.   A multivariate analysis of
       wilderness users in the United States. In #682
       (Krutilla), pp. 142-170.

Presents results of a survey of users of four wilderness areas, to show relationships between selected demographic variables, as well as the user's outdoor experience as a child, and diverse aspects of the user's orientation towards wilderness (e.g., commitment, response to crowding, mode of travel, etc.).

697. Cicchetti, C. J. A review of the empirical analyses that have been based upon the National Recreation Surveys. Journal of Leisure Research1, 1972, 4, 90-107.

A review of the surveys of outdoor recreation conducted by the Outdoor Recreation Resources Review Commission, including data on preferences for and participation in different activities and reasons given for these preferences and choices.

698. Cicchetti, C. J., & Smith, V. K. Congestion, quality deterioration, and optimal use: Wilderness recreation in the Spanish Peaks Primitive Area. Social Science Research, 1973, 2, 15-30.

A model for the analysis of the impact of congestion on the wilderness experience. Data on willingness-to-pay to preserve solitude are presented, and their use for the formulation of management policy discussed.

699. Clark, R. N., Hendee, J. C. & Campbell, E. L. Values, behavior, and conflict in modern camping culture. Journal of Leisure Research, 1971, 3, 143-159.

A comparative study of recreation managers and campers with regard to attitudes and values concerning the camping experience, and maintenance-services and management issues.

700. Driver, B. L. Potential contributions of psychology
        to recreation resource management. In #14
        (Wohlwill & Carson), pp. 233-244.

    A general treatment of motivational and social aspects
of natural recreation, and of the contribution of psychology
to the recreation manager.

701. Driver, B. L., & Knopf, R. C. Temporary escape -- one
        product of sport fisheries management. Fisheries,
        1976, 1(2), 24-29.

    Data from surveys of fishermen are presented to document
the part played by the opportunity for temporary escape from
everyday sources of stress in the environment, and of change
of pace from everyday living, as a major factor motivating
the fisherman. The role of other factors (development of
skill, strengthening of family ties) is also discussed.

702. Driver, B. L., & Knopf, R. C. Personality, outdoor
        recreation and expected consequences. Environment
        and Behavior, 1977, 9, 169-193.

    An exploratory study of relationships between
personality (measured by Jackson's Personality Research
Form) and choice of and frequency of participation in
selected recreation activities, and benefits expected from
them.

703. Ferriss, A. L. The social and personality correlates
        of outdoor recreation. In S. Klausner (Ed.),
        Society and its physical environment. Annals of
        the American Academy of Political and Social
        Science, 1970, #389, 46-55.

    A discussion of personality and socioeconomic factors
related to participation in different forms of physical
recreation, including athletics and natural recreation

activities. Trends over time are presented over the period 1960 to 1965.

704. Gilbert, C. G., Peterson, G. L., & Lime, D. W. Toward a model of travel behavior in the Boundary Waters Canoe area. Environment and Behavior, 1973, 4, 131-157.

Describes a stochastic model, based on Markov renewal theory, for the analysis of movement patterns of wilderness canoers.

705. Glock, C. Y., Sleznick, G., & Wiseman, T. The wilderness vacationist. In WILDERNESS AND RECREATION: A REPORT ON RESOURCES, VALUES AND PROBLEMS. (Outdoor Recreation Resources Review Commission Study Report #3). Washington, D. C.: U. S. Government Printing Office, 1962. Pp. 126-202.

Presents an extensive survey of the wilderness user, including demographic characteristics, data on correlated childhood experience, urban vs. rural residence, characteristics of wilderness trips, bases for the appeal of wilderness, and attitudes towards wilderness management.

706. Greist, D. A. The carrying capacity of public wild land recreation areas: Evaluation of alternative measures. Journal of Leisure Research, 1976, 8, 123-128.

Compares two approaches to the conceptualization of the carrying capacity of a natural recreation area, one based on the use level at which satisfaction is maximized, the other based on use level demanded after considering costs. Measurement considerations lead to a preference for the latter approach.

707. Hancock, H. K. Recreation preference: Its relation
     to user behavior. Journal of Forestry, 1973, 71,
     336-337.

     Compared campers' verbally expressed preferences for
amount of vegetation in the vicinity of campsites with the
actual amount of use of campsite areas systematically
subjected to progressive reduction in trees and ground cover
vegetation. Preferences favored maximal amounts of
vegetation, yet use increased as vegetation was removed up
to 80%.

708. Heberlein, T. A. Density, crowding, and satisfaction:
     Sociological studies for determining carrying
     capacity. In #684 (U. S. Forest Service, 1977),
     pp. 67-76.

     An analysis of the concept of carrying capacity,
differentiating among physical, ecological, facilities, and
social aspects. Suggests a social-norms view of carrying
capacity, in preference to a satisfaction model favored by
natural-resource economists.

709. Heberlein, T. A., & Shelby, B. Carrying capacity,
     values and the satisfaction model: A reply to
     Greist. Journal of Leisure Research, 1977, 9,
     142-148.

     A discussion of issues of level of measurement in
measures of carrying capacity based on aggregate
satisfaction, as well as of conceptual issues relating to
Greist's satisfaction model (see #706).

710. Hendee, J. C. Rural-urban differences reflected in
     outdoor recreation participation. Journal of
     Leisure Research, 1970, 2, 333-341.

An examination of differences between rates of participation in rural as opposed to urban populations in outdoor recreation, in terms of contrasting hypotheses emphasizing compensatory processes, familiarity, opportunity, etc.

711.   Hendee, J. C., & Burgess, R. L.   The substitutability concept: Implications for recreation research and management. Journal of Leisure Research, 1974, 6, 157-162.

A discussion of the substitutability concept as used in leisure research, defined as the interchangeability of different recreation activities in satisfying needs, and motives of the recreationist. Implications for recreation management are considered.

712.   Hendee, J. C., Catton, W. R., Jr., Marlow, L. D., & Brockman, C. F. Wilderness users in the Pacific Northwest—their characteristics, values, and management preferences. (U. S. Forest Service Research Paper PNW—61). Portland, Oregon: Pacific Northwest Forest and Range Experiment Station, 1968. 92 pp.

An investigation of demographic, attitudinal and behavioral characteristics of users of different wilderness areas of the Northwest. The Wildernism-Urbanism scale is used to determine correlates of attitudes towards wilderness.

713.   Hendee, J. C., Gale, R., & Catton, W. R. A typology of outdoor recreation activity preferences. Journal of Environmental Education, 1971, 3(1), 28-34.

Presents a typology of preferred activities, in terms of five categories: Appreciative-symbolic, extractive-symbolic, passive free-play, sociable learning and active-

expressive. Examples of each are given, and relationships
to age and education suggested.

714. Hendee, J. C., & Harris, R. W. Forester's perception
       of wilderness-user attitudes and preferences.
       Journal of Forestry, 1970, 68, 759-762.

   Compared attitudes of wilderness users regarding diverse
aspects of their wilderness experience and alternative
policies for wilderness management to the attitudes of
wilderness managers concerning the same issues, as well as
the latter's perceptions of the attitudes and views of
users. Important discrepancies between user attitudes and
managers' perceptions of them are brought out.

715. Kaplan, R. Some psychological benefits of an outdoor
       challenge program. Environment and Behavior,
       1974, 6, 101-116.

   Ten participants in a wilderness survival program were
studied before, immediately after and several months after
the program. Certain long-term benefits were suggested by
the results.

716. Kellert, S. R. Perceptions of animals in American
       society. Transactions, 41st North American
       Wildlife and Natural Resources Conference, 1976,
       533-546.

   Develops a typology of orientations towards animals
consisting of nine attitudes (e.g., ecologistic, humanistic,
scientistic, utilitarian, etc.), and reports results of a
survey relating this typology to the person's dominant mode
of relating to animals (e.g., as pet owner, hunter, breeder,
etc.), and to demographic variables.

717. Kellert, S. R. Attitudes and characteristics of
       hunters and antihunters. Transactions, 43d North

American Wildlife and Natural Resources
Conference, 1978, 412-423.

A study comparing groups of hunters and of persons
opposed to hunting in terms of their standing on nine types
of attitudinal orientation towards animals. The two groups
were most sharply differentiated along the dimensions of
utilitarian and dominionistic (i.e., master-and control-
oriented) attitudes, favoring the hunters, and on a
moralistic dimension, favoring the anti-hunters.

718. Kennedy, J. J. Attitudes and behavior of deer hunters
     in a Maryland forest. Journal of Wildlife
     Management, 1974, 38, 1-8.

Results from a mailed questionnaire survey show that
hunters have a relatively high tolerance for density of
other hunters in their vicinity. Companionship and
enjoyment of the out-of-doors loomed more important than
success in bagging game.

719. Knopf, R. C., Driver, B. L., & Bassett, J. B.
     Motivations for fishing. Transactions of the 38th
     North American Wildlife and Natural Resources
     Conference, 1973, 191-204.

A report of exploratory research on the motivational
basis for fishing, describing the development of a measuring
instrument, and some preliminary results based on
comparisons among diverse groups of recreationists,
including two small samples of fishermen.

720. Knopp, T. B. Environmental determinants of recreation
     behavior. Journal of Leisure Research, 1972, 4,
     129-138.

Emphasizes the role of environmental factors, as opposed
to the particular activities engaged in, as determinants of

satisfaction from recreational activity. Argues for a
compensatory model of choice of environments for recreation,
and reports data from a comparison of rural and urban
residents of a midwestern county, supporting this model.

721. Lee, R. G. Alone with others: the paradox of privacy
       in wilderness. Leisure Sciences, 1977, 1, 3-20.

       Questionnaire-based attitudes of hikers in the Yosemite
backcountry toward crowding and privacy on the trails were
found to bear little relationship to their social
interaction responses to other persons encountered on the
trail, or to choice of campsites.

722. Lime, D. W. Behavioral research in outdoor recreation
       management: An example of how visitors select
       campgrounds. In #14 (Wohlwill & Carson), pp.
       198-206.

       Report of a study of the environmental factors
influencing the selection of campgrounds, with implications
for concerns of the recreation manager.

723. Lucas, R. C. Wilderness perception and use: The
       example of the Boundary Waters Canoe Area.
       Natural Resources Journal, 1964, 3, 394-411.

       A report of an investigation of managers' and users'
perception of wilderness, in terms of basis for users'
choice of campsites, criteria used in deciding on the area
considered as wilderness, and response to encounters with
other users. Relationships between perception and actual
use are brought out.

724. McNeil, E. B. The background of therapeutic camping.
       In # 725 (McNeil), pp. 3-14.

A review of research and thinking on the use of camping
experiences with children for psychotherapeutic purposes.
Widely shared assumptions concerning the value of camping
for this purpose contrast with a lack of solid evidence
demonstrating beneficial effects.

725. McNeil, E. B. (Ed.), Therapeutic camping for disturbed
        youth. Journal of Social Issues, 1957, 13(1), 62
        pp.

A collection of papers on therapeutic camping, including
discussions of the use of space and behavioral ecology of
campers, therapeutic use of social isolation, and case
studies of special camping programs.

726. Mercer, D. The role of perception in the recreation
        experience: A review and discussion. Journal of
        Leisure Research, 1971, 3, 261-276.

A brief review of perceptual, attitudinal and behavioral
aspects of outdoor recreation, organized in terms of
Clawson's five-stage model of the recreation experience:
anticipation, travel to the site, on-site activity, return
travel, and recollection.

727. Mercer, D. The concept of recreational need. Journal
        of Leisure Research, 1973, 5, 37-50.

A general discussion of the concept of recreational
need, arguing for a pluralistic conception of need, and
differentiating between normative, comparative, felt and
expressed needs.

728. Mercer, D. C. Motivational and social aspects of
        recreational behavior. In #8 (Altman & Wohlwill,
        1), pp. 123-162.

A review of perceptual, motivational and social-
psychological research on recreational behavior, emphasizing
outdoor recreation, and examining the relationship between
residential and preferred recreational environments, the
role of personality factors, and social forces.

729. Moeller, G. H., & Engelken, J. H. What fishermen look
      for in a fishing experience. Journal of Wildlife
      Management, 1972, 36, 1253-1257.

Results from a survey of 100 fishermen (including both
lake and stream fishing) are cited to bring out the
important place of characteristics of the environment and of
the fisherman's experience in it (water quality, natural
beauty, privacy) as prime determinants of the enjoyment
users derive from their activity.

730. More, T. A. Attitudes of Massachusetts hunters.
      Transactions of the 38th North American Wildlife
      and Natural Resources Conference, 1973, 203-234.

A survey of hunters, based on a mail-sample of 325
questionnaires, suggests that their motivation for hunting
resembles that of participants in other forms of outdoor
recreation, notably in regard to the role of aesthetic
factors and of the element of challenge; the kill itself
emerges as of subsidiary importance.

731. Nielsen, J. C., & Endo, R. Where have all the purists
      gone? An empirical examination of the
      displacement hypothesis in wilderness recreation.
      Western Sociological Review, 1977, 8, 61-75.

A survey of persons who have taken a raft trip through
the Grand Canyon over a period of 17 years, to determine
whether a significant proportion of wilderness
recreationists move from areas of high use and crowding to
less crowded ones. The results showed little evidence of
such a displacement phenomenon.

732. Peterson, G. L. A comparison of the sentiments and
        perceptions of wilderness managers and canoeists
        in the Boundary Waters Canoe Area. Journal of
        Leisure Research, 1974, 6, 194-206.

        A comparison of managers and canoeists revealed basic
similarity in environmental dispositions (using McKechnie's
ERI, see item 98), along with differences in specific
evaluative and preference responses and in images and
knowledge of the area.

733. Peterson, G. L. Evaluating the quality of the
        wilderness environment; Congruence between
        perception and aspiration. Environment and
        Behavior, 1974, 6, 169-193.

        Presents a quantitative model for assessing wilderness
quality, through the determination of the degree of
congruence between the rated desirability of each of a large
number of specific characteristics and the extent to which
the environment is perceived as containing that
characteristic, leading to an overall index of satisfaction.

734. Peterson, G. L., & Lime, D. W. Two sources of bias in
        the measurement of human response to the
        wilderness environment. Journal of Leisure
        Research, 1973, 5 , 66-73.

        Results of mail surveys of wilderness users are
criticized in terms of the role of two potential sources of
bias: (1) users' retrospective evaluations of their
experience when responding to their survey may not be valid
indicators of their experience while in the field; (2) there
are differences between those who do and those who do not
return questionnaires.

735. Potter, D. R., Hendee, J. C., & Clark, R. N., Hunting
        satisfaction: Game, guns, or nature?

Transactions of the 38th North American Wildlife
and Natural Resources Conference, 1973, 220-229.

A questionnaire survey of a large sample of hunters in
the state of Washington was used to determine major
dimensions of hunter satisfaction. Results showed 11
different dimensions, of which the most important were
enjoyment of nature, escapism and change of pace, and
companionship. Bagging game emerged as relatively
subsidiary in importance.

736. Rossman, B. B., & Ulehla, J. Psychological reward
      values associated with wilderness use: A
      functional-reinforcement approach. Environment
      and Behavior, 1977, 9, 41-66.

An in-depth study of the rewards wilderness users expect
to obtain from wilderness areas, stressing the importance of
emotional and aesthetic factors, and of change of pace from
urban life. The substitution value of alternative
recreational settings decreased in proportion to their
dissimilarity to a wilderness setting.

737. Schreyer, R., & Roggenbuck, J. W. The influence of
      experience expectations on crowding perceptions
      and social-psychological carrying capacities.
      Leisure Sciences, 1978, 1, 373-394.

A study of river recreationists' perception of crowding
in relation to their expectations concerning the
recreational experience, and attitudes towards wilderness.

738. Scott, N. R. Toward a psychology of wilderness
      experiences. Natural Resources Journal, 1974, 14,
      231-237.

A discussion of possible self-actualization and related
values of wilderness experience.

739.   Shafer, E. L. Jr., & Burke, H. D.   Preferences for
        outdoor recreation facilities in four state parks.
        Journal of Forestry, 1965, 63, 512-518.

    A comparison of campers vs. non-campers in regard to
their preferences for differently designed and planned
facilities (e.g., swimming areas, picnic areas, etc.). The
alternative facilities were presented via photographs, with
different hypothetical prices associated with each.

740.   Shafer, E. L., Jr., & Mietz, J.   Aesthetic and
        emotional experiences rate high with northeast
        wilderness hikers. Environment and Behavior,
        1969, 1 187-197.

    Backpackers hiking in the Adirondack Mountains were
asked to rank-order the relative importance of different
aspects of their recreation experience. Results pointed to
the importance of aesthetic and emotional aspects, while
educational and social values played a more subsidiary role.
That of physical exercise was intermediate in importance.

741.   Shafer, E. L., Jr., & Moeller, G.   Predicting
        quantitative and qualitative values of recreation
        participation. In #683 (Larson), pp. 5-22.

    Discusses approaches to the development of predictive
equations for rate of recreation participation of use to
recreation managers. A constellation of variables,
including distance, recreation-supply, socio-economic
characteristics, and qualitative measures of the
recreational environment, is proposed as serving this
purpose effectively.

742.   Shafer, E. L., Jr., & Thompson, R. C.   Models that
        describe use of Adirondack campgrounds. Forest
        Science, 1968, 14, 383-391.

Dimensions of variation in the attributes of campgrounds
are determined via a factor analysis of 40 site
characteristics of 20 campgrounds, and related to data on
campground use.

743.   Stankey G. H. Wilderness: Carrying capacity and
       quality. Naturalist, 1971, 22(3), 7-13.

A critical analysis of the concept of carrying capacity,
with particular reference to the perception of carrying
capacity by wilderness users. Results from a questionnaire
study are reported.

744.   Stankey, G. H. A strategy for the definition and
       management of wilderness quality. In #682
       (Krutilla), pp. 88-114.

A discussion of wilderness quality in terms of user
perception, satisfaction and behavior, including
consideration of the definition of wilderness by the user,
the concept of carrying capacity, and aspects of visitor
satisfaction.

745.   Stone, G. P., & Taves, M. J. Camping in the
       wilderness. In E. Larrabee and Z. Meyersohn
       (Eds.), MASS LEISURE. Glencoe, Ill.: Free Press,
       1958. Pp. 290-304.

A review of social-scientific aspects of camping,
including demographic characteristics of campers, wilderness
images held by campers, and issues in the sociology of
camping.

746.   Turner, A. L. The therapeutic value of nature.
       Journal of Operational Psychiatry, 1976, 7, 64-74.

A review of research on beneficial effects of camping and survival-training experiences on self-concept and personality functioning in adolescents and mental patients.

747. Twight, B. W., & Catton, W. R., Jr., The politics of images: Forest managers vs. recreation publics. Natural Resources Journal, 1975, 15, 297-306.

A general discussion of "unshared images," i.e., discrepancies between the assumptions and perspectives of natural resource managers and those of the users of the resource. Includes a brief report of a study of visitors to an arboretum, pointing to the importance of the appearance of the landscape, and overall atmosphere, as opposed to the specific horticultural and educational values of the exhibits.

748. Wohlwill, J. F., & Heft, H. A comparative study of user attitudes towards development and facilities in two contrasting natural recreation areas. Journal of Leisure Research, 1977, 9, 264-280.

In two separate populations--campers in state parks, and lessees of State Forest sites in Pennsylvania--more favorable attitudes towards development and facilities, and less wilderness-oriented attitudes were found among persons frequenting an area relatively high in development as compared to those in a more isolated, undeveloped area.

749. Yoesting, D., & Burkead, D. Significance of childhood recreation experience on adult leisure behavior: An exploratory analysis. Journal of Leisure Research, 1973, 5, 25-36.

Report of a study of the relationship between participation in diverse recreation activities in childhood and as an adult, for a sample of residents of rural counties in Iowa. Significant relationships were found for 14 of 35 activities.

750.   Yoesting, D. R., & Christensen, J. E. Reexamining the
       significance of childhood recreation patterns on
       adult leisure behavior. Leisure Sciences, 1978,
       1, 219-230.

       Demonstrated a relationship between extent of
participation in outdoor recreation activities in childhood
and in adulthood; this relationship was general, rather than
activity-specific, holding for total amount of
participation, but for only eight of 45 specific activities.

751.   Al-Hoory, M. T. Campsite selection as a decision-
       making process: A behavioral approach.
       Dissertation Abstracts International, 1976, 37-A,
       3173.

752.   Buhyoff, G. J. The use of behavioral measurements to
       assess on-site recreation preferences.
       Dissertation Abstracts International, 1975, 36-B,
       1117-1118.

753.   Grubb, E. A. Temporal integration as a measure of job
       boredom and consequent selection of recreation
       activities. Dissertation Abstracts International,
       1973, 34-B, 1722-1723.

754.   Hendee, J. C. Recreation clientele--the attributes of
       recreationists preference. Dissertation
       Abstracts, 1967, 28-B, 1747.

755.   Roggenbuck, J. W. Socio-psychological inputs into
       carrying capacity assessments for float-trip use
       of white-water rivers in Dinosaur National
       Monument. Dissertation Abstracts International,
       1976, 37-A, 3173.

756. Shaw, W. W. Sociological and psychological
        determinants of attitudes towards hunting.
        Dissertation Abstracts International, 1974, 35-B,
        5471.

757. Shelby, B. B. Social-psychological effects of
        crowding in wilderness: The case of river trips
        in the Grand Canyon. Dissertation Abstracts
        International, 1977, 37-B, 4227-4228.

758. Smith, M. A. W. An investigation of effects of an
        outward-bound experience on selected personality
        factors and behavior of high-school juniors.
        Dissertation Abstracts International, 1971, 32-A,
        6141.

759. Stankey, G. H. The perception of wilderness
        recreation carrying capacity. Dissertation
        Abstracts International, 1971, 32-B, 1655-1656.

760. Stribling, J. C. A comparison of methods of
        characterizing recreationists and recreation
        behavior. Dissertation Abstracts International,
        1977, 37-A, 8002-8003.

761. Wetmore, R. C. The influence of an outward bound
        school experience on the self-concept of
        adolescent boys. Dissertation Abstracts
        International. 1972, 33-A, 1498-1499.

See also: citations 524, 610, 680, 813, 862, 915.

d. Natural Hazards

762. Burton, I., Kates, R., & White, G. THE ENVIRONMENT AS
      HAZARD. New York: Oxford University Press, 1978.
      xvi, 240 pp.

      An overview of the problem of natural hazards in its
various manifestations across the globe, with particular
reference to drought, floods, cyclones and earthquakes.
Modes of individual choice and collective action are
considered, and implications for national policy and
international action discussed.

763. Kates, R. W., RISK ASSESSMENT OF ENVIRONMENTAL HAZARD
      New York: Wiley, 1978. (SCOPE Project 8) xvii,
      112 pp.

      A general overview. Includes material on methodology of
risk assessment, and on attitudes in assessing environmental
threats from both natural hazards and technology.

764. White, G. (Ed.), NATURAL HAZARDS: LOCAL, NATIONAL,
      GLOBAL. New York: Oxford University Press, 1974.
      xvi, 288 pp.

      A collection of reports dealing with diverse hazards in
specific local contexts throughout the world; most include a
description of the hazard and its past history of
occurrence, and data on perceptual, behavioral and societal
response to it.

765. Adams, R. L. A. Uncertainty in nature, cognitive
      dissonance, and the perceptual distortion of
      environmental information: Weather forecasts and
      New England beach trip decisions. Economic
      Geography, 1973, 49, 287-297. [reprinted in #33
      (Sims & Baumann), pp. 162-178.]

A questionnaire study of the influence of weather-forecast information on decisions to carry out or cancel (hypothetical) plans to go to the beach, revealing some distortions in the assessment of the threat of unfavorable weather in accordance with previously made plans.

766. Burton, I. Cultural and personality variables in the
      perception of natural hazards. In #14 (Wohlwill &
      Carson), pp. 184-195.

A brief review of research on cultural and personality determinants of the response to natural hazards on the part of both individuals and society. Differentiates between preindustrial, industrial and postindustrial pattern of response, based respectively on individual and small-group adjustments, technological intervention, and a combination of technological and behavioral mechanisms.

767. Burton, I., & Kates, R. W. The perception of natural
      hazards in resource management. Natural Resources
      Journal, 1964, 3, 412-441.

A discussion of the differing ways in which scientists, resource managers and resource users (e.g., farmers) view natural hazards.

768. Golant, S., & Burton, I. A semantic differential
      experiment in the interpretation and grouping of
      environmental hazards. Geographical Analysis,
      1970, 2, 120-134. [Reprinted in #136 (Moore &
      Golledge), pp. 364-374].

A factor-analytic study of the connotative dimensions of meaning attached to natural and man-made hazards

769. Hanson, S., Vitek, J. D., & Hanson, P. O. Natural
      disaster: long-range impact on human response to
      future disaster threats. Environment and
      Behavior, 1979, 11, 268-284.

A study of the perception of and response to the hazard
of tornados on the part of residents of a mid-Eastern city
that had experienced a major tornado twenty-five years
earlier. Awareness of that event was found to predict
readiness to take adequate precautions against future
tornado threat.

770. Kates, R. W. HAZARD AND CHOICE PERCEPTION IN FLOOD
        PLAIN MANAGEMENT. Chicago: University of Chicago
        Press, 1962. (Dept. of Geography Research Paper
        #78). 157 pp.

A systematic study of the perception of hazard from
floods on the part of residents of various localities,
relating variations in such perception to actual frequency
of occurrence of floods at each locality, and reporting
differences in the perceived patterning of the recurrence of
floods within and among localities.

771. Kates, R. W. The perception of storm hazard on the
        shores of megalopolis. In D. Lowenthal (Ed.),
        ENVIRONMENTAL PERCEPTION AND BEHAVIOR (Department
        of Geography Research Paper 109). Chicago:
        University of Chicago, 1967. Pp. 60-74.

An account of a questionnaire-based study of the
assessment of the hazards associated with residence in a
storm-prone area along the Atlantic coast, with particular
reference to the perceived regularity and predictability of
such events.

772. Kates, R. W. Natural hazard in human ecological
        perspective: Hypotheses and models. Economic
        Geography, 1971, 47, 438-451.

Develops a general systems model of human adjustment to
natural hazards, taking account of hazard-perception

thresholds, personal experience and evaluation and search processes.

773. Kates, R. W. Experiencing the environment as hazard. In #54 (Wapner et al.), pp. 133-156.

Presents a cognitive model of the experience of the environment as hazard. Differentiates between natural, man-made and social-environmental hazards, as well as between different consequences from hazards and different modes of coping with them.

774. Saarinen, T. F. PERCEPTION OF DROUGHT HAZARD ON THE GREAT PLAINS. Chicago: University of Chicago, Department of Geography, 1966. (Research Paper #106). 183 pp.

A study of Midwestern farmers' assessment of the probability of drought occurrence, showing an overall tendency to underestimate drought hazard, but greater sensitivity to this hazard on the part of farmers in areas most prone to droughts.

775. Schiff, M. Hazard adjustment, locus of control and sensation seeking: Some null findings. Environment and Behavior, 1978, 9, 233-254.

No relationship was found between verbal measures of adjustment to prospective hazards and measures of individual differences in either locus of control or sensation-seeking tendency. Adjustment to natural and man-made hazards were found to be closely related.

776. Sims, J. H., & Baumann, D. D. The tornado threat: Coping styles of the North and South. Science, 1972, 176, 1386-1392. [Reprinted in #33 (Sims & Baumann), pp. 108-125].

Suggests a psychological interpretation of the observed
higher death rates from tornados in the South as compared to
the North, based on a more fatalistic attitude assumed to
characterize Southerners.

777.   Slovic, P. Assessment of risk taking behavior.
       Psychological Bulletin, 1964, 61, 220-233.

A general review of research on risk-taking, and on
individual differences in risk-taking behavior, emphasizing
its multi-dimensional character and raising questions
concerning the validity of risk-taking as a construct.

778.   Slovic, P., Kunreuther, H., & White, G. F. Decision
       processes, rationality, and adjustment to natural
       hazards. In G. F. White (ed.), NATURAL HAZARDS:
       LOCAL, NATIONAL, GLOBAL. New York: Oxford
       University Press, 1974. Pp. 187-205.

An overview of decision theories and of Simon's concept
of "bounded rationality" and a review of research findings
on probabilistic information processing and risk taking
behavior in the laboratory, as well as of field studies of
response to natural hazards. This research evidence is
interpreted within the framework of the "bounded
rationality" concept.

779.   Sorensen, J. H., & White, G. F. Natural hazards: A
       cross-cultural perspective. In #11 (Altman et.
       al., 4), pp. 279-318.

A general review of natural-hazard research, with
emphasis on cross-national and inter-cultural comparisons.
Two models, a general-systems model and an economic
development model, are applied to the literature in this
area; the role of cognitive and decision-making processes in
interpreting differential response to hazards is also
stressed.

780. Van Arsdol, M. D., Jr., Sabagh, G., & Alexander, F.
Reality and the perception of environmental
hazards. Journal of Health and Human Behavior,
1964, 5, 144-153.

A survey of actual and perceived presence of
environmental hazards in the Greater Los Angeles area, from
smog, aircraft noise, brush fire, earth slide and flood,
noting differential relationships between reality and
perception, and the role of the forms of urbanization in
that area.

781. Lucas, R. Z. Social Behavior under conditions of
extreme stress: A study of miners entrapped by a
coal mine disaster. Dissertation Abstracts
International, 1971, 31-A, 4906.

782. Sorensen, J. H. Interaction of adjustments to natural
hazard. Dissertation Abstracts International,
1977, 38-A, 3052.

8. ATTITUDINAL AND BEHAVIORAL ASPECTS OF
ENVIRONMENTAL
PROBLEMS

Psychology has not been at the forefront of the disciplines that have converged on the analysis and on the alleviation of the pressing problems of environmental pollution, conservation, and quality more generally. Yet the study of environmental attitudes represents a natural subject of inquiry for social psychologists interested in the psychology of attitudes, and in such classic problems as the relationships between attitudes and behavior. There has been a scattering of studies by psychologists in this field, along with a considerably larger body of attitudinal literature contributed by sociologists (and rural sociologists in particular), as well as certain other disciplines, such as political science.

While the analysis of the role of behavior as a contributor to environmental problems might seem to represent an obvious concern for psychology, it should be recognized that it entails a 180 degree reversal in the typical $B = f(E)$ formula implicit or explicit in almost all psychological research; that is, the conceptualization of behavior as a function of environmental variables. The possibility that environments might in turn be affected by human behavior has remained of limited interest to psychologists qua psychologists. The signal exception is that of a group following the behavior-modification approach who have sensed here an opportunity for the application of behavior-modification principles to the amelioration of urgent practical problems facing society. The work of this group represents the bulk of the material included in the second subsection of this chapter, dealing with behavioral components of environmental issues. Attention should also be drawn, however, to very recent contributions from social psychology which represents a very different point of view (cf. items #864 and #878).

a. Environmental Attitudes

783.   Althoff, P., & Grieg, W. H. Environmental pollution
          control policymaking: An analysis of elite
          perceptions and preferences. Environment and
          Behavior, 1974, 6, 259-288.

     A comparison of attitudes towards environmental controls
held by leaders in business and industry, elected officials
and governmental officials.

784.   Arbuthnot, J., & Lingg, S. A comparison of French and
          American environmental behaviors, knowledge, and
          attitudes. International Journal of Psychology,
          1975, 10, 275-281.

     The interrelationship between knowledge, attitude and
behavior with respect to recycling is compared in French and
American adults.

785.   Asch, J., & Shore, B. M. Conservation behavior as the
          outcome of environmental education. Journal of
          Environmental Education, 1975, 6(4), 25-33.

     Demonstrated changes in a small group of fifth-grade
boys taking part in a two-year environmental-education
program in the direction of environment-conserving as
opposed to environment-destroying forms of behavior (e.g.,
with respect to conservation of water and soils, protection
of wildlife and forests, etc.).

786.   Barker, M. L. Information and complexity: The
          conceptualization of air pollution by specialist
          groups. Environment and Behavior, 1974, 6,
          346-377.

A comparison of students receiving training in five
fields (law, medicine, engineering, economics and geography)
in terms of their definition, knowledge and concern for air
pollution, their assessment of environmental information,
and their view of their own professional role in this area.

787. Bruvold, W. H. Consistency among attitudes, beliefs,
       and behavior. Journal of Social Psychology, 1972,
       86, 127-134.

Examined attitude-belief and attitude-behavior
relationships concerning the use of recycled water.

788. Bruvold, W. H. Belief and behavior as determinants of
       environmental attitudes. Environment and
       Behavior, 1973, 5, 202-218.

A further study of the interrelationship between
attitudes, overt behavior and beliefs, concerning the use of
recycled water. Attitudes were found to be consistently
related to belief on the one hand and behavior on the other,
and to be co-determined by them.

789. Buttel, F. H., & Flinn, W. L. The structure of
       support for the environmental movement, 1968-1970.
       Rural Sociology, 1974, 39, 55-69.

Survey data collected over a three-year period disclose
a spread of environmental concern over a progressively
broader segment of the population, though not extending to
the working class.

790. Buttel, F. H., & Flinn, W. L. Conceptions of rural
       life and environmental concern. Rural Sociology,
       1977, 42, 544-555.

Suggests a distinction between agrarianism and ruralism,
the former referring to values specifically related to
farming, the latter referring to rural as opposed to urban
life-styles. In a study of Wisconsin residents, these two
sets of values did not correlate differentially with
environmental concern.

791. Buttel, F. H., & Flinn, W. L. The politics of
          environmental concern: The impacts of party
          identification and political ideology on
          environmental attitudes. Environment and
          Behavior, 1978, 10, 17-36.

A statewide survey of Wisconsin residents showed
environmental attitudes to be related to sociopolitical
ideology (notably liberalism vs. "laissez-faire"
conservatism), but only weakly to political-party
identification.

792. Buttell, F. H., & Johnson, D. E. Dimensions of
          environmental concern: Factor structure,
          correlates and implications for research. Journal
          of Environmental Education, 1977, 9(2), 49-64.

Differentiated between amelioration and redirection as
alternative orientations towards the solution of
environmental problems, the latter involving advocacy of
major institutional and societal changes. Measures of these
two orientations were obtained, and correlations of each
with selected demographic and attitudinal variables are
reported.

793. Constantini, E., & Hanf, K. Environmental concern and
          political elites: A study of perceptions,
          backgrounds, and attitudes. Environment and
          Behavior, 1972, 4, 209-242.

A comparison of different decision-making groups in the
Lake Tahoe region with respect to their concern for
environmental quality in that region.

794. Coughlin, R. E. The perception and valuation of water
      quality: A review of research method and
      findings. In #238 (Craik & Zube), pp. 205-228.

      A review of research on perception of and attitude
toward water quality, including perception of water
pollution, and its relation to objective indices of
pollution, to attitude, and to use.

795. Creer, R. N., Gray, R. M., & Teshow, M. Differential
      responses to air pollution as an environmental
      problem. Journal of the Air Pollution Control
      Association, 1970, 20, 814-818.

      Persons economically dependent on a source of pollution
profess less concern with it than those not so dependent.
Results are interpreted in terms of cognitive dissonance
theory.

796. DeGroot, I. Trends in public attitudes toward air
      pollution. Journal of the Air Pollution Control
      Association, 1967, 17, 679-681.

      A brief survey of research on awareness of and attitudes
towards air pollution, disclosing consistent relationships
between actual amount of pollution and awareness and
concern. These relationships were found to have definite
limitations.

797. Devall, W. B. Conservation: An upper-middle class
      social movement: A replication. Journal of
      Leisure Research, 1970, 2, 123-126.

      A survey of Sierra Club members generally supports the
contention of Harry et al. (#811) concerning the upper-
middle class make-up of environmentalist organization

membership, but found more evidence of generalism, i.e.,
participation by such members in a diversity of other
voluntary organizations.

798. Dillman, D. A., & Christenson, J. A. The public value
        for pollution control. In #1449 (Burch, Cheek &
        Taylor), pp. 237-256.

A demographic analysis of factors related to favorable
attitudes towards pollution control.

799. Dispoto, R. G. Interrelationships among measures of
        environmental activity, emotionality and
        knowledge. Educational and Psychological
        Measurement, 1977, 37, 451-459.

Found a positive relationship between knowledge about
ecology, pollution and the environment, and involvement in
activities related to environmental problems.

800. Donohue, G. A., Olien, C. N., & Tichenor, P. J.
        Communities, pollution, and fight for survival.
        Journal of Environmental Education, 1974, 6(1),
        29-37.

A follow-up study to Tichenor et al. (#842) comparing
sentiment on diverse environmental issues and knowledge
concerning specific problems on the part of residents of
four Minnesota communities, showing changes occurring over a
two-year interval.

801. Downs, A. Up and down with ecology -- the "issue-
        attention" cycle. The Public Interest, 1972, 28
        (Summer), 38-50.

Suggests that shifts in public concern over an issue
such as environmental quality goes through a five-phase
cycle involving a waxing and waning of concern and interest.

802. Dunlap, R. E. The impact of political orientation on
     environmental attitudes and actions. Environment
     and Behavior, 1975, 7, 428-454.

   Among a sample of college students studied in 1970,
various dimensions of concern for environmental issues and
views on their solution were found to be correlated with
both political preference and the respondent's standing on a
liberalism-conservatism scale.

803. Dunlap, R. E., & Dillman, D. A. Decline in public
     support for environmental protection: Evidence
     from a 1970-1974 panel study. Rural Sociology,
     1976, 41, 382-390.

   Found a sharp decline, between 1970 and 1974, in the
priority given to both general and specific environmental
problems seen as warranting the expenditure of tax dollars.

804. Dunlap, R. E., & Heffernan, W. B. Outdoor recreation
     and environmental concern: An empirical
     examination. Rural Sociology, 1975, 40, 1975,
     18-30.

   Differentiates between an "appreciative" and a
"consumptive" orientation towards outdoor recreation. Data
from a survey of Washington State residents suggest that the
former is more conducive to environmental concern than the
latter.

805. Dunlap, R. E., & Van Liere, K. D. Land ethic or
     golden rule. Comment on "Land ethic realized" by
     Thomas A. Heberlein. Journal of Social Issues,
     1977, 33(3), 200-207.

A critique of Heberlein's article (see #812) which
argued that a change in ethical values has taken place in
people's overall orientation towards the environment. The
authors contend that the change is a largely temporary one,
and in any event based not so much on ethical considerations
as on recognition of the necessity of measures for combating
pollution and upgrading environmental quality for the
protection of human health.

806.   Dunlap, R. E., & Van Liere, K. D.   The "new
          environmental paradigm." Journal of Environmental
          Education, 1978, 9(4), 10-19.

     Responses to a 12-item questionnaire dealing with
fundamental concepts of environmental quality, such as
"balance of nature," "limits of growth," "steady-state
society," are cited to show a large measure of awareness of
and concern for environmental quality on the part of an
unselected sample of the population, leading to the
suggestion of a "new environmental paradigm." Data on the
validity and internal consistency of the 12-item scale are
presented.

807.   Erskine, H.   The polls: Pollution and its costs.
          Public Opinion Quarterly, 1972, 36, 120-135.

     A compilation of data from major public opinion polls
(Gallup, Harris, Roper) taken during the period from 1965 to
1971, dealing with issues of pollution and environmental
quality.

808.   Geisler, C. C., Martinson, O. B., & Wilkening, E. A.
          Outdoor recreation and environmental concern: A
          restudy. Rural Sociology, 1977, 42, 241-249.

     A reexamination of the hypothesis of Dunlap and
Heffernan (#804) suggesting that environmental concern is
related to an "appreciative" rather than a "consumptive"
orientation towards outdoor recreation. Results from a

large sample of Wisconsin residents showed that this
association vanished, once effects of age, education and
place of residence were partialled out.

809.   Goodman, R. F., & Clary, B. B.   Community attitudes
       and action in response to airport noise.
       Environment and Behavior, 1976, 8, 441-470.

A survey of residents in several communities in the
vicinity of Los Angeles International Airport assessed the
extent of disturbance to various activities, from reading
(least disturbed) to TV watching (most affected); individual
and community political response to the problem is also
discussed.

810.   Hardison, M., & Vanier, D. J.   Environment and quality
       of life: Attitudinal measurement and analysis.
       Psychological Reports, 1976, 39, 959-965.

A survey of attitudes toward pollution and quality of
life in San Diego, Calif.

811.   Harry, J., Gale, R. P., & Hendee, J.   Conservation:
       An upper-middle class movement. Journal of
       Leisure Research, 1969, 1, 246-254.

Members of an outdoor recreation organization were
divided into "conservationists" -- those who belonged to a
conservation organization -- and 'nonconservationists' --
those who did not. The former were older, more highly
educated and of higher socio-economic status than the
latter.

812.   Heberlein, T. A.   The land ethic realized: Some
       social psychological explanations for changing
       environmental attitudes. Journal of Social
       Issues, 1972, 28(4), 79-87.

On the basis of experimental socio-psychological
research, an interpretation is offered for the change from
an economic to a moral orientation toward environmental
issues.

813. Heberlein, T. A. Social psychological assumptions of
       user attitude surveys: The case of the Wildernism
       scale. Journal of Leisure Research, 1973, 5,
       18-33.

    A discussion of attitude theory illustrated with respect
to attitudes towards wilderness. Includes the horizontal
and vertical structuring of attitudes, the relationship
between attitudes and behavior, and issues of attitude
change.

814. Heberlein, T. A., & Black J. S. Attitudinal
       specificity and the prediction of behavior in a
       field setting. Journal of Personality and Social
       Psychology, 1976, 32, 474-479.

    Found that an environmentally relevant form of behavior
-- purchase of lead-free gasoline -- was more strongly
related to beliefs relating specifically to the role of
gasoline in air pollution than to more generic measures of
environmental concern.

815. Hetrick, C. C., Lieberman, C. J., & Ranish, D. R.
       Public opinion and the environment: Ecology, the
       coastal zone and public policy. Coastal Zone
       Management Journal, 1974, 1, 275-289.

    A survey of residents of Santa Barbara county regarding
their attitudes on coastal zone development and regulation,
and related environmental issues. Attitudes were found to
be closely related to age, education and extent of political
participation. A general high level of environmental
concern in this population was not accompanied by
willingness to allocate a larger share of individual income
to the solution of environmental problems.

816. Hohm, C. F. A human-ecological approach to the
      reality and perception of air pollution. Pacific
      Sociological Review, 1976, 19, 21-44.

An examination of diverse demographic and other factors
modulating the relationship between actual and perceived
severity of air pollution in the Los Angeles area.

817. Hummel, C. F., Levitt, L., & Loomis, R. J.
      Perceptions of the energy crisis: Who is blamed
      and how do citizens react to environment-life
      style trade-offs? Environment and Behavior, 1978,
      10, 37-88.

A study of residents of a Colorado community, during and
after the 1973 gasoline shortage, in terms of support for
voluntary and mandatory measures for energy conservation and
pollution-abatement, and for measures involving a conflict
between energy and pollution concerns.

818. Humphrey, C. R., Bord, R. J., Hammond, M. M., & Mann,
      S. H. Attitudes and conditions for cooperation in
      a paper recycling program. Environment and
      Behavior, 1977, 9, 107-124.

A study of environmental conditions and attitudinal
factors related to office workers' receptivity to
participation in a wastepaper recycling program and to their
accuracy in separating paper from other waste materials.

819. Jorgenson, D. O. Measurement of desire for control of
      the physical environment. Psychological Reports,
      1978, 42, 603-608.

Reports on the development and validation of a 14-item
scale to measure the individual's orientation towards the

environment and nature, in terms of the strength of the
desire to achieve control over it.

820. Koenig, D. J.   Additional research on environmental
        activism. Environment and Behavior, 1975, 7,
        472-485.

    Studied correlates of environmental activism, which was
found to be related to general social activism and
liberalism, but not to education and income.

821. Krauss, R. M., Freedman, J. L., & Whitcup, M.   Field
        and laboratory studies of littering. Journal of
        Experimental Social Psychology, 1978, 14, 109-122.

    Via a combined laboratory and field research
methodology, a consistent relationship was found between an
individual's tendency to engage in littering and the amount
of litter present in an area.

822. Kromm, D. E., Probald, F., & Wall, G.   An
        international comparison of response to air
        pollution. Journal of Environmental Management,
        1973, 1, 363-375.

    Compared respondents in three European cities (in
England, Hungary and Yugoslavia) with respect to their
awareness of air pollution, their ability to identify its
chemical components, the behavioral adjustments made to
pollution, and the assessment of future trends.

823. Levenson, H.   Perception of environmental
        modifiability and involvement in antipollution
        activities. Journal of Psychology, 1973, 84,
        237-239.

Members of a local antipollution organization were found to show less optimism regarding the tractability of environmental problems than were non-members, reflecting an apparently greater reluctance to opt for the "technological fix" in handling problems of pollution.

824. Lipsey, M. W. Attitudes toward the environment and pollution. In S. Oskamp (Ed.), ATTITUDES AND OPINIONS. Englewood Cliffs, N. J.: Prentice-Hall, 1977. Pp. 360-379.

A review of research on demographic variables related to environmental concern, influences mediating such concern, participation in pro-environment activities, and attitudes relating to the solution of environmental problems.

825. Lounsbury, J. W., & Tornatzky, L. G. A scale for assessing attitudes towards environmental quality. Journal of Social Psychology, 1977, 101, 299-305.

Developed a three-dimensional scale including concern for environmental degradation, environmental action, and overpopulation. These differentiated members of environmental-action organizations from non-members and predicted overall levels of behavioral involvement in environmental issues.

826. Maloney, M. P., Ward, M. P., & Braucht, G. N. A revised scale for the measurement of ecological attitudes and knowledge. American Psychologist, 1975, 30, 787-790.

Presents a scale for the measurement of environmental attitudes, with four subscales: Verbal commitment, actual commitment, affect and knowledge. Data on the validation of the scale when administered to Sierra Club members, college students and non-college adults are reported.

827. McEvoy, J. III The American concern with environment.
      In #1449 (Burch, Cheek & Taylor), pp. 214-236.

    An analysis of demographic correlates of environmental
concern.

828. McGuinness, J., Jones, A. P., & Cole, S. G.
         Attitudinal correlates of recycling behavior.
         Journal of Applied Psychology, 1977, 62, 376-384.

    A study of participation in a paper-recycling program by
residents of different socio-economic levels, and its
relationship to expressed ecological attitudes.

829. Medalia, N. Z. Air pollution as a socio-environmental
         health problem. Journal of Health and Human
         Behavior, 1964, 5, 154-165.

    A study of residents of a small town living in the
vicinity of a pulp mill. Concern with pollution was found
positively related to involvement in the community. Concern
and perceived seriousness were also higher among those who
had resided in town for more than 10 years as compared with
more recent arrivals.

830. Milbrath, L. W., & Sahr, R. C. Perceptions of
         environmental quality. Social Indicators
         Research, 1975, 1, 397-438.

    A comparative survey of environmental satisfaction based
on ratings of evaluation and importance of diverse
attributes of the environment held by urban and rural
residents of New York State. Possibilities and problems in
the construction of an index of perceived environmental
quality are discussed.

831. Miles, J. C. The study of values in environmental
       education. Journal of Environmental Education,
       1977, 9(2), 5-17.

A discussion of concepts of value as utilized in
anthropological and psychological theory (Kluckhohn;
Kohlberg's theory of moral judgment) and their application
to environmental values.

832. O'Riordan, T. Public opinion and environmental
       quality: A reappraisal. Environment and
       Behavior, 1971, 3, 191-214.

An analysis of the structure of public opinion as it
relates to environmental problems with particular reference
to local issues, and of group conflicts arising in the
environmental-decision making process.

833. O'Riordan, T. Attitudes, behavior, and environmental
       policy issues. In #8 (Altman & Wohlwill 1), pp.
       1-36.

Applies a cognitive model to the understanding of
environmental attitudes and their relation to behavior.
Implications for environmental policy and environmental
education are discussed.

834. Pallak, M. S., & Cummings, W. Commitment and
       voluntary energy conservation. Personality and
       Social Psychology Bulletin, 1976, 2, 27-30.

In a study conducted at the height of the 1973 energy
crisis, a public commitment to energy conservation was found
to result in lower natural gas and electricity use than
either a private commitment or no commitment.

835. Pierce, J. C. Water resource preservation: Personal
      values and public support. Environment and
      Behavior, 1979, 11, 147-161.

     A survey of about 700 residents of the State of
Washington, to determine value correlates of attitudes
towards water resource preservation, using Rockeach's
inventory of personal values. Different aspects of water-
preservation attitudes related differentially to the values
of "a world of beauty" vs. that of "a comfortable life;"
differences between waterfront property owners and others
are noted.

836. Rankin, R. D. Air pollution control and public
      apathy. Journal of the Air Pollution Control
      Association, 1969, 19, 565-569.

     A survey of Charleston, W. Va. residents confirming
previous studies in showing respondents to be more apt to
attribute pollution as a problem to their city as a whole
than to their own neighborhood, regardless of the actual
severity of pollution in their area.

837. Reid, D. H., Luyben, P. L., Rawers, R. J., & Bailey,
      J. S. Newspaper recycling behavior: The effects
      of prompting and proximity of container.
      Environment and Behavior, 1976, 8, 471-482.

     A study of participation in a newspaper-recycling
program by residents of apartment complexes.

838. Robinson, S. N. Littering behavior in public places.
      Environment and Behavior, 1976, 8, 363-384.

     A review of littering behavior, examining its
relationship to environmental, demographic and cognitive-
personality variables.

839. Schaeffer, D. J., & Janardan, K. G. Communicating
      environmental information to the public: A new
      water quality index. Journal of Environmental
      Education, 1977, 8(4), 18-26.

     The authors review various indices proposed previously
as measures of water quality, and propose one devised by
themselves. An application of the index to a sample of
observation posts disclosed good agreement of the resulting
values with direct ratings of water quality on a four-point
scale.

840. Sharma, N. C., Kivlin, J. E., & Fliegel, F. C.
      Environmental pollution: Is there enough public
      concern to lead to action. Environment and
      Behavior, 1975, 7, 455-471.

     An analysis of reactions to a local water pollution
problem as well as determinants of individual involvement in
such issues, and impediments to attempts at promoting
involvement.

841. Simpson, C. J., Rosenthal, T. L., Daniel, T. C., &
      White, G. M. Social-influence variations in
      evaluating managed and unmanaged forest areas.
      Journal of Applied Psychology, 1976, 61, 759-763.

     Studied the influence of social anchoring techniques and
social persuasion techniques on acceptability ratings of
slides of natural, partially thinned, and clear-cut forest
scenes. Judgements of the latter two types of scenes were
susceptible to both types of influence.

842. Tichenor, P. J., Donohue, G. A., Olien, C. N., &
      Bower, J. K. Environment and public opinion. The
      Journal of Environmental Education, 1971, 2(4),
      38-42.

A comparison of knowledge of and sentiment on diverse
environmental issues in four Minnesota communities. (cf.
also Donohue et al., #800).

843. Tognacci, L. N., Weigel, R. H., Wideen, M. F., &
       Vernon, D. T. A. Environmental quality: How
       universal is public concern? Environment and
       Behavior, 1972, 4, 73-86.

A study of University town residents shows their
attitudes on environmental quality to be closely related to
broader sentiment and views on issues of social welfare, and
general socio-political outlook. These results lead the
authors to question the breadth of support for the
environmental movement among the rank and file.

844. Tucker, L. R. The environmentally concerned citizen:
       Some correlates. Environment and Behavior, 1978,
       10, 389-418.

An investigation of relationships between attitudinal
and behavioral measures of environmental responsibility, and
measures of general social responsibility, internal-external
control, and demographic variables.

845. Wall, G. Public response to air pollution in South
       Yorkshire, England. Environment and Behavior,
       1973, 5, 219-248.

Public opinion regarding air pollution generally, and
towards a legislative measure to combat it was found to be
unrelated to a given community's adoption or rejection of
that legislation; the latter depended rather on the actions
and influence of special interest groups.

846. Weigel, R. H. Ideological and demographic correlates
       of proecology behavior. Journal of Social
       Psychology, 1977, 103, 39-47.

A sample of 91 residents of a medium-sized New England
town disclosed relationships between environmental attitudes
and educational and occupational status and social and
political liberalism, confirming results from prior
research. In a subsample of 44 respondents, the same
relationship was found to hold for actual participation in
ecology projects.

847. Weigel, R. H., Vernon, D. T. A., & Tognacci, L. N.
        Specificity of the attitude as a determinant of
        attitude-behavior congruence. Journal of
        Personality and Social Psychology, 1974, 30,
        724-728.

Behavioral commitment to participation in Sierra Club
activities was related to varying degree of specificity of
environmental attitudes, from high specificity (attitude
toward the Sierra Club) to low specificity (attitude towards
achieving a pure environment). The hypothesis that
attitude-behavior congruence is proportional to specificity
of this relationship was confirmed.

848. Weigel, R., & Weigel, J. Environmental concern: The
        development of a measure. Environment and
        Behavior, 1978, 10, 3-16.

The authors developed a 16-item "Environmental Concern
Scale," showing good internal consistency, test-retest
reliability, and validity in differentiating Sierra Club
members from non-members. Scores on this scale also
correlate highly with participation in environmentally
relevant activities.

849. White, G. F. Formation and role of public attitudes.
        In H. Jarrett (Ed.), ENVIRONMENTAL QUALITY IN A
        GROWING ECONOMY. Baltimore: John Hopkins
        University Press, 1966. Pp. 105-127.

A general treatment of issues and methods in the study of environmental attitudes and values.

850. Wicker, A. W. Attitudes versus actions: The relationships of verbal and overt behavioral responses to attitude objects. Journal of Social Issues, 1969, 25 (4), 41-78.

An integrative review of literature on the relationships between expressed attitudes and overt behaviors, with reference to diverse social and political issues.

851. Wohlwill, J. F. The social and political matrix of environmental attitudes: An analysis of the vote on the California Coastal Zone Regulation Act. Environment and Behavior, 1979, 11, 71-86.

On the basis of a precinct level analysis of the vote in the 1972 California election, sentiment in favor of a measure to control development of the coastal zone was found to be correlated in varying degrees with other propositions dealing with issues of social and moral philosophy as well as another environmental issue relating to anti-pollution control. Relationship to the Presidential vote, and to several demographic variables, was low.

852. Baca, T. P. A study of the environmental attitudes of four different age groups. Dissertation Abstracts International, 1977, 37-A, 7555-7556.

853. Goquiolay, S. T. An empirical investigation of ecologically responsible consumers and their buying behavior. Dissertation Abstracts International, 1977, 37-A, 6021-6022.

854. Kameron, J. B. Man-nature value orientations. Dissertation Abstracts International, 1975, 35-B, 6076.

855. Pinsley, A. A. The greening of the environment: A
      study of cognitions which predispose people to
      join environmental groups. Dissertation Abstracts
      International, 1975, 36-B, 1119-1120.

856. Steinberg, H. W. Relationships between level of
      personality development and individuals' value
      orientations. Dissertation Abstracts
      International, 1977, 37-B, 5445-5446.

See also: citations 239, 244, 344, 663, 716, 717, 860, 885,
1414, 1426.

b. The Role of Behavior, and Behavior Change, in
Environmental Problems

857. Cone, J. D., & Hayes S. C. Applied behavior analysis
      and the solution of environmental problems. In #9
      (Altman & Wohlwill, 2), pp. 129-180.

      A comprehensive review of empirical research and theory
on the application of behavior-modification approaches to
human behavior that adversely affect resource conservation
and environmental health. Specific problems considered
include littering, noise, recycling, transportation, and
energy use.

858. Arbuthnot, J. et al., The induction of sustained
      recycling behavior through the foot-in-the-door
      technique. Journal of Environmental Systems,
      1977, 6, 355-368.

This paper reports an application of an attitude-
behavior change technique, based upon assumed changes in
self-perception, to recycling behavior. The method involves
the use of a series of relatively simple and non-demanding
appeals which culminate in an appeal for the desired
behavior, i. e. use of a community recycling center.

859. Arbuthnot, J. The roles of attitudinal and
      personality variables in the prediction of
      environmental behavior and knowledge. Environment
      and Behavior, 1977, 9, 217-232.

Compared users of a recycling center with members of a
conservative church congregation, in regard to their
participation and involvement in environmental concerns,
their knowledge about and sources of information on
environmental issues, and their responses to environmental
attitude and personality scales. Differential correlates of
knowledge as opposed to behavior are pointed out.

860. Bickman, L. Environmental attitudes and actions.
      Journal of Social Psychology, 1972, 87, 323-324.

In two separate experiments, littering behavior was
found (a) not to be influenced by observation of a model
picking up litter, and (b) unrelated to verbally expressed
attitudes concerning littering.

861. Chapman, C., & Risley, T. R. Anti-litter procedures
      in an urban high-density area. Journal of Applied
      Behavior Analysis, 1974, 7, 377-383.

The problem of trash and litter in a low-income public
housing project for blacks was attacked by paying children
(aged 4 to 13) either to collect trash or to keep an
assigned yard area clean and free of trash. The latter
procedure resulted in a marked alleviation of the problem.

862. Clark, R. N., Burgess, R. L., & Hendee, J. C. The development of antilitter behavior in a forest campground. Journal of Applied Behavior Analysis, 1972, 5, 1-5.

Report of a study in which children aged 6 to 14 were successfully enlisted to pick up campground litter for small rewards such as a shoulder patch or a ruler.

863. Deslauriers, B. C., & Everett, P. B. The effects of intermittent and continuous token reinforcement on bus ridership. Journal of Applied Psychology. 1977, 62, 369-375.

Ridership of a campus bus increased to an equivalent degree over baseline through the application of either continuous-reinforcement (receipt of a small token by every passenger) or variable-reinforcement (receipt of a token by every third passenger) procedures.

864. Edney, J. J., & Harper, C. S. The effects of information in a resource management problem: A social trap analog. Human Ecology, 1978, 6, 387-395.

Utilized a gaming technique to study the mode of response of members of a group to a situation involving an exhaustible resource pool. Results pointed to the general tendency to adopt a short-term gain strategy, even at the expense of accelerated depletion of the resource.

865. Everett, P. B., Hayward, S. C., & Meyers, A. W. The effects of a token reinforcement procedure on bus ridership. Journal of Applied Behavior Analysis, 1974, 7, 1-9.

Use of a campus bus service was substantially increased through disbursement to riders of tokens exchangable for diverse articles of merchandise.

866.   Finnie, W. C.   Field experiments in litter control.
       Environment and Behavior, 1973, 5, 123-144.

       An account of several field studies of littering
behavior in urban areas, pointing to the important role of
environmental stimuli (e.g., availability and design of
litter receptacles) in controlling such behavior.

867.   Foxx, R. M., & Hake, D. F.   Gasoline conservation: A
       procedure for measuring and reducing the driving
       of college students. Journal of Applied Behavior
       Analysis, 1977, 10, 60-74.

       Demonstrated the effectiveness of different monetary,
material and other incentives for voluntary reduction of
driving behavior among college students. Driving decreased
by an average of 20 percent relative to baseline over a one-
month period.

868.   Geller, E. S., Chaffee, J. L., & Ingram, R. E.
       Promoting paper recycling on a university campus.
       Journal of Environmental Systems, 1975, 5, 39-57.

       Newspaper recycling by dormitory residents was increased
both through a lottery, and through a contest among
different dormitories, with a cash prize to the winning
dormitory. The former, individual, condition proved more
effective than the latter, communal one.

869.   Geller, E. S., Farris, J. C., & Post, D. S. Prompting
       a consumer behavior for pollution control Journal
       of Applied Behavior Analysis, 1973, 6, 367-376.

       Studied the effectiveness of handbills urging the
purchase of returnable bottles distributed in a store, and
of charts recording each such a purchase as it was made, in

increasing the purchase rate of soft drinks in returnable
bottles.

870. Hayes, S. C., & Cone, J. D. Decelerating
          environmentally destructive lawn-walking behavior.
          Environment and Behavior, 1977, 9, 511-534.

    A demonstration of the effectiveness of different forms
of environmental control (e.g., through signs, physical
barriers, etc.) in curbing lawn-walking behavior on a
university campus.

871. Jordan, P. A. Ecology, conservation, and human
          behavior. Bioscience, 1968, 18, 1023-1029.

    A biological scientist's view of environmental problems,
and of the role of human behavior both as contributors to
them and in the devising of solutions to them.

872. Kohlenberg, R., Phillips, T., & Proctor, W. A
          behavioral analysis of peaking in residential
          electrical energy consumption. Journal of Applied
          Behavior Analysis, 1976, 9, 13-18.

    Compared the effectiveness of information concerning the
problem of peaking in energy consumption, feedback in regard
to actual peaking in a given household, and monetary
incentives for reduction in peaking. A combination of
feedback and incentives proved most effective.

873. Mazis, M. B., Settle, Z. B., & Leslie, D. C.
          Elimination of phosphate detergents and
          psychological reactance. Journal of Marketing
          Research, 1973, 10, 390-395.

    An investigation of response to enforced elimination of
phosphate detergents through a city ordinance. Housewives

in the city of Miami were compared with a control group for
which phosphate use remained legal. Persons forced to
switch brands were compared with those who were not.
Results are interpreted in terms of psychological reactance
theory.

874. Meyers, A. W., Artz, L. M., & Craighead, W. E. The
        effects of instructions, incentives, and feedback
        on a community problem: Dormitory noise. Journal
        of Applied Behavior Analysis, 1977, 9, 445-447.

        Noise levels on a dormitory floor were reduced through a
combination of instruction concerning the learned-behavior
basis of excessive noise as well as feedback (ringing of a
bell whenever a threshold noise level was exceeded). More
dramatic reductions were achieved by adding reinforcement,
in the form of either monetary rewards or academic grades,
to reduction in noise on the floor.

875. Seaver, W. B., & Patterson, A. H. Decreasing fuel-
        consumption through feedback and social
        commendation. Journal of Applied Behavior
        Analysis, 1976, 9, 147-152.

        In a study conducted during the oil-crisis of 1974, home
fuel consumption was significantly reduced through a
combination of feedback and commendation. Feedback alone
was ineffective.

876. Seligman, C., & Darley, J. M. Feedback as a means of
        reducing residential energy consumption. Journal
        of Applied Psychology, 1977, 62, 363-368.

        Demonstrated the effectiveness of feedback provided
daily, in reducing electricity use (primarily air-
conditioning), during a three-week summer period.

877. Stern, P. C. Effect of incentives and education on resource conservation decisions in a simulated commons dilemma. Journal of Personality and Social Psychology, 1976, 34, 1285-1292.

A gaming-simulation study of car-pooling behavior, indicating the role of negative incentives of price increases in promoting car pooling; direct payoffs and gasoline rationing proved ineffective. Limitations on the generalizability of these results to "real-world" behavior are considered.

878. Stern, P. C., & Kirkpatrick, E. M. Energy behavior. Environment, 1977, 19(9), 10-15.

An analysis of psychological aspects of energy consumption, and of the use of incentives to effect a reduction in such consumption, examining the relevance of principles of learning and of reactance theory to this problem.

879. Tuso, M. A., & Geller, E. S. Behavior analysis applied to environmental-ecological problems: A review. Journal of Applied Behavior Analysis, 1976, 9, 526.

An abstract of a review of behavior-modification approaches to littering, lawn-walking, recycling, and energy consumption, in terms of discriminative stimuli and reinforcement procedures. (Complete paper available from National Auxiliary Publication Service.)

880. Winett, R. A. Efforts to disseminate a behavioral approach to energy conservation. Professional Psychology, 1976, 7, 222-228.

Describes a demonstration project on the effectiveness of monetary incentives in reducing electrical energy use,

and discusses the obstacles to acceptance and application of
such behavioral research on the part of public utilities and
similar organizations.

See also: citations 818, 837, 894, 895.

9. BEHAVIORAL ASPECTS OF TRANSPORTATION, TRAVEL AND TRAFFIC

While problems of transportation could be considered to represent a specific environmental issue, and could thus have been included in the previous chapter, several reasons dictate a separate chapter for this topic. First, the field includes not only work on attitudes and behavior change strategies related to transportation use, but also research on the psychological impacts of travel and traffic, and other responses to transportation environments, considered as a particular type of setting. Thus the topic of transportation serves as a bridge between the preceding chapter and those to follow, which deal with specific environmental settings. Furthermore, the behavioral study of transportation has become identified as a special area of applied psychology; it encompasses the efforts of specialists in human factors, in economic behavior and decision theory, and others in related areas, and has thus become differentiated from both the study of environmental problems in general and from that of other specific environmental settings.

881. Altman, I., Wohlwill, J. F., & Everett, P. B. (Eds.). HUMAN BEHAVIOR AND ENVIRONMENT, V: TRANSPORTATION ENVIRONMENTS. New York: Plenum, in preparation. (To appear in 1981).

Papers in this volume will include such topics as transportation demand modeling, approaches to behavior change related to transportation use, behavioral aspects of transportation safety, transportation as a source of stress, social impacts of transportation systems, and models of transportation for the future.

882. Forbes, T. W. (Ed.). HUMAN FACTORS IN HIGHWAY TRAFFIC SAFETY RESEARCH. New York: Wiley-Interscience, 1972. xix, 419 pp.

Includes chapters on driver simulation studies, highway signs and other information systems, the pedestrian, driver fatigue, effects of alcohol and drugs on driving behavior, and others.

883. Shepard, M. E. Annotated bibliography of the application of behavior and attitude research to transportation system planning and design. Monticello, Ill.: Council of Planning Librarians, 1976. (Exchange Bibliographies, #1066) 31 pp.

884. Transportation Research Board, National Research Council. Behavioral demand modeling and valuation of travel time. Special Report, 1974, #149. 234 pp.

Proceedings of a Conference conducted by the Highway Research Board.

885. Amir, S., Highway location and public opposition. Environment and Behavior, 1972, 4, 413-436.

A case study of public and community opposition to new highway construction, focused on the history of the Hudson River Expressway in New York State, examining nature and bases for opposition, forms of protest strategy, and the dynamics of conflict between citizens groups and decision-making agencies.

886. Benshoof, J. A. Characteristics of drivers' route selection behaviour. Traffic Engineering and Control, 1970, 11, 604-606.

Found that some of the reasons stated by drivers for their choice of one among alternative routes were based on a distortion of the objective situation, notably in regard to choice of shortest route and of route with least traffic.

887. Brancher, D. Traffic in scenic areas: The user-road relationship. Journal of the Town-Planning Institute, 1968, 54, 271-274.

A brief sketch of a model for conceptualizing relationships between the character of a road (rural-urban), user volume, user type (environmentally sensitive-insensitive) and user behavior, with implications for the design and management of roads.

888. Brand, D. Approaches to travel behavior research. Transportation Research Record, 1976, #569, 12-33.

An overview of issues and findings in research on travel behavior, with emphasis on choice processes and the isolation of determinants of travel choice.

889. Bronzaft, A. L., Dubrow, S. B., & O'Hanlon, T. J. Spatial orientation in a subway system. Environment and Behavior, 1976, 8, 575-594.

A study of subway orientation and trip planning problems of subjects from New York City in executing prescribed trips on the New York subways. Limitations of the official Subway Guide, and of maps and graphics provided in the subway are noted.

889a. Brown, I. D. & Poulton, E. C. Measuring the spare "mental capacity" of car drivers by a subsidiary task. Ergonomics, 1961, 4, 35-40.

While driving a car through either a residential or
shopping area, subjects listened to and then indicated
changes in subsequent sets of eight digit numbers. Errors
in shopping areas, presumably as a consequence of more
competing inputs, than in the residential areas.

890.   Burkhardt, J. E., & Shaffer, M. T.   Social and
       psychological impacts of transportation
       improvements. Transportation, 1978, 7, 207-226.

A theoretical analysis of the impacts of transportation
facilities at the individual and community levels.

891.   Demetsky, J., & Hoel, L.   Modal demand: A user
       perception model. Tranportation Research, 1972,
       6, 293-308.

A model for the analysis of user perceptions and modal
choice behavior in regard to transportation.

892.   Dobson, R. Towards the analysis of attitudinal and
       behavior responses to transportation system
       characteristics. Transportation, 1975, 4,
       267-290.

Application of general behavior theory, emphasizing
reinforcement principles, to analysis of behavioral
consequences of transportation systems and changes in such
systems.

893.   Dobson, R., Golob, T. F., & Gustafson, R. L.
       Multidimensional scaling of consumer preferences
       for a public transportation system: An
       application of two approaches. Socio-Economic
       Planning Science, 1974, 8, 23-36.

An application of two multidimensional scaling
techniques (a vector model vs. an unfolding model) to
people's preferences for a new transportation system.

894. Everett, P. B., & Hayward, S. C. Behavioral
         technology: An essential design component of
         transportation systems. High Speed Ground
         Transportation Journal, 1974, 8, 139-143.

A presentation of the principles of behavior theory and
the role of reinforcement in application to transportation
behavior, illustrated with findings from a study in which
behavior-modification principles were used to change rates
of bus ridership on a college campus.

895. Everett, P. B., Studer, R. G., & Douglas, T. J.
         Gaming simulation to pretest operant-based
         community interventions: An urban transportation
         example. American Journal of Community
         Psychology, 1978, 6, 327-338.

Developed an urban transportation game to simulate bus-
riding behavior and determine the effectiveness of different
schedules of reinforcement.

896. Ewing, R. H. Psychological theory applied to mode
         choice prediction. Transportation, 1973, 2,
         391-410.

Application of a psychophysical model to the prediction
of use of a dial-a-bus system in a new town. Results point
to the need for individual rather than aggregate data, and
to the role of adaptation-level in modulating choices of
transportation mode.

897. Golob, T. F. Resource paper on attitudinal models.
         Highway Research Board, National Research Council,
         Special Report, 1973, #143, pp. 130-145.

A review of the psychology of attitudes and of attitude-behavior relations, and their application to transportation behavior.

898.   Golob, T. F., Canty, E. T., Gustafson, R. L., & Vitt, J. E. An analysis of consumer preferences for a public transportation system. Transportation Research, 1972, 6, 81-102.

A description of a broad-based market research survey of urban residents' preference for public transportation systems.

899.   Gustafson, R. L., Curd, H. N., & Golob, T. F. User preferences for a demand-responsive transportation system: A case study report. Highway Research Record, 1971, #367, 31-45.

A survey of preferences for various demand-responsive transportation systems (such as dial-a-ride) in a suburban population. Results showed the importance of level of service, convenience, and costs.

900.   Hartgen, D. T. Attitudinal and situational variables influencing urban mode choice: Some empirical findings. Transportation, 1974, 3, 377-392.

Choice of transportation mode was related to various situational and attitudinal variables. The former, including auto-ownership and income, outweighed the latter.

901.   Hartgen, D. T., & Tanner, G. T. Individual attitudes and family activities: A behavioral model of travel choice. High Speed Transportation Journal, 1970, 4, 439-467.

An analysis of individual choices of travel mode based
on a framework of individual behavior and attitude in the
context of household activities and family roles.

902. Hartgen, D. T., & Tanner, G. T. Investigations of the
      effect of traveler attitudes in a model of mode-
      choice behavior. Highway Research Record, 1971,
      #369, 1-14.

Description and text of a model for predicting choice of
transportation mode from respondents' attitudes towards
specified attributes of a given mode.

903. Hille, S. J., & Martin, T. K. Consumer preference in
      transportation. Highway Research Record, 1967,
      #197, 36-43.

A factor analytic study of a survey of residents of
Baltimore and surrounding rural areas, to determine the
importance of different attributes of a transportation
system. Six major factors emerged that were independent of
trip-purpose: lack of repairs, travel time, cost,
independence from control, traffic and age of vehicle.
Additionally, reliability was high in importance for trips
to work or school.

904. Hoag, L. L., & Adams, S. K. Human factors in urban
      transportation systems. Human Factors, 1975, 17,
      119-131.

A discussion of vehicular design and other aspects of
transportation systems in terms of considerations of
anatomical, psychomotor and general behavior requirements of
different user groups.

905. Hornbeck, P. L., Forster, R. R., & Dillingham, M. R.
      Highway aesthetics--functional criteria for
      planning and design. Highway Research Record,
      1969, #280, 25-38.

A general discussion of aesthetic considerations in highway design and construction, including both visual and behavioral aspects. A demonstration case study is described.

906. Horton, F., & Louviere, J. Behavior analysis and
      transit planning: Inputs to transit planning.
      Transportation, 1974, 3, 165-182.

Distinguishes between data on feelings, attitudes, and opinions considered important to assessment of system performance, and data on disposition to respond or use considered essential in decisions relating to manipulation of system parameters. Use of the former to monitor user satisfaction before and after introduction of an innovation in a bus system is described.

907. Kalfon, C., Yardon, W., & Menkes, J. Measurement of
      users' preferences for public transportation
      through computer assisted models. Transportation
      Science, 1975, 9, 21-32.

Preferences for different bus systems by residents of Boulder, Colorado were studied by a gaming approach in which subjects had to design a system by assigning given values to different characteristics, and validated with data on estimates of probability of use of a given system.

908. Kinley, H. J. Resource paper on the transportation
      disadvantaged. Highway Research Board, National
      Research Council, Special Report, 1973, #143,
      121-129.

A survey of transportation behavior and needs in specific "disadvantaged" groups, e.g., car-less and car deficient households, the elderly, and non-whites.

909. LeBoulanger, H. Research into the urban traveller's
        behaviour. Transportation Research, 1971, 5,
        113-125.

    A theoretical analysis of travel behavior, based in
large part on French transportation research.

910. Louviere, J. J., & Norman, K. L. Applications of
        information-processing theory to the analysis of
        urban travel demand. Environment and Behavior,
        1977, 9, 91-106.

    A theory of the integration of multiple forms of
information is applied to subjects' evaluations of
hypothetical bus-systems on the basis of specifications of
diverse characteristics, e.g., fares, travel time, walking
distance, etc.

911. Michaels, R. M. Attitudes of drivers determine choice
        between alternative highways. Public Roads, 1965,
        33, 225-236.

    A survey of over 3,000 drivers to determine the bases
for their preference between two alternative routes: a toll
expressway vs. a primary rural highway. The results point
to the role of stress and stress-management as major
factors, overriding the effects of cost and driving-time.

912. Michaels, R. M. Behavioral measurement: An approach
        to predicting transport demand. Transportation
        Research Board, National Research Council, Special
        Report, 1974, #149, 51-57.

    A brief treatment of measurement and scaling aspects of
research on transportation preference and demand.

913.   Michon, J. A.   The mutual impacts of transportation
         and human behaviour.   In P. Stringer & H. Wenzel
         (Eds.), TRANSPORTATION PLANNING FOR A BETTER
         ENVIRONMENT.   New York: Plenum, 1976. Pp.
         221-235.

      A general overview of relations between transportation
and behavior.

914.   Mourant, R. R., & Rockwell, T. H.   Mapping eye-
         movement patterns to the visual scene in driving:
         An exploratory study.   Human Factors, 1970, 12,
         81-87.

      Eye movements, under either open-road or car-following
conditions, were monitored for eight drivers using an urban
expressway.   Changes in eye-movements as the driver's
familiarity with the road increased were observed.

915.   O'Rourke, B.   Travel in the recreational experience--
         a literature review.   Journal of Leisure Research,
         1974, 6, 140-156.

      A review of research on travel (particularly by
automobile) as a component of recreational activity, with
emphasis on choice of mode of travel, of destination, and
other behavioral factors.

916.   Paine, F. T., Nash, A. N., & Hille, S. J.   Consumer
         attitudes toward auto versus public transport
         alternatives.   Journal of Applied Psychology,
         1969, 53, 472-480.

      Reports results of two surveys carried out in Baltimore
and Philadelphia showing the influence of specific
attributes of private vs. public modes of transportation on
preferences for one or the other.

917. Quarmby, D. A. Choice of travel mode for the journey
to work: Some findings. Journal of Transport
Economics and Policy, 1967, 1, 273-314.

A study of commuting by car vs. public transport in
Leeds England, with reference to perception of trade-offs
between time and cost.

918. Recker, W. W., & Golub, T. F. An attitudinal modal
choice model. Transportation Research, 1976, 10,
299-310.

A model for predicting choice of mode of travel for
trips to work on the basis of selected mode attributes.

919. Reichman, S. Subjective time-savings in interurban
travel: An empirical study. Highway Research
Record, 1973, #446, 22-17.

In a study of air, bus, and private car passengers in
Israel direct estimates of amount of time saved in going by
air were compared with estimates of amount of time spent for
a given trip in going by each of the three modes. A strong
tendency to overestimate amount of time saved was found,
along with a strongly bi-modal distribution of the time-
saving estimates, which is related to characteristics of the
various trips.

920. Singer, J. E., Lundberg, U., & Frankenhaeuser, M.
Stress on the train: A study of urban commuting.
In #12 (Baum, Singer & Valins), pp. 41-56.

Verbal as well as physiological measures were obtained
from commuter-train passengers, and related to length of
trip, to determine reactions of discomfort and stress.

921. Stopher, P. On the application of psychological
     measurement techniques to travel demand
     estimation. Environment and Behavior, 1977, 9,
     67-80.

     A discussion of the application of methods of
psychological scaling techniques to determine the relative
attractiveness of alternative travel destinations and
alternative modes of travel. The application of market-
segmentation techniques in defining homogeneous subgroups of
the population in travel research is also considered.

922. Stopher P. R., Spear, B. D., & Sucher, P. O. Toward
     the development of measures of convenience for
     travel models. Highway Research Record, 1974,
     #527, 16-32.

     Application of scaling techniques to the study of the
role of convenience in choice of transportation mode.

923. Ulrich, R. S. SCENERY AND THE SHOPPING TRIP: THE
     ROADSIDE ENVIRONMENT AS A FACTOR IN ROUTE CHOICE.
     Ann Arbor: University of Michigan, Dept. of
     Geography. 1974. xvii, 176 pp. (Michigan
     Geographical Publication #12).

     Studied the bases for drivers' selection of an
expressway as opposed to a route through a suburban area in
driving to a shopping center, in terms of the respective
roles of aesthetic appeal and other aspects of the route,
time and other factors.

924. Wachs, M. Relationships between drivers' attitudes
     towards alternative routes and driver and route
     characteristics. Highway Research Record, 1967,
     #197, 70-87.

A survey of residents of Evanston, Ill. concerning their preference for alternate driving routes. These preferences are related to characteristics of drivers and routes.

925. Wildervanck, C., Mulder, G., & Michon, J. A. Mapping mental load in car driving. Ergonomics, 1978, 21, 225-229.

A methodological note describing a system for the study of mental load in driving, utilizing a videotape recorder to monitor traffic and physiological and performance indices of mental load.

926. Marks, M. E. Individual differences in transportation behaviors, Dissertation Abstracts International, 1979, 39-B, 4107.

927. McGhee J. L. Public choice: Transportation preferences of community decision makers and transportation consumers. Dissertation Abstracts International, 1978, 39-A, 3879.

See also: citations 118, 298, 300, 333, 555, 863, 865, 867.

10. INSTITUTIONAL SETTINGS

This chapter focuses on a variety of social institutions, all of which share several important characteristics: all serve specific populations (e.g., students, elderly, inmates); all are intended to serve particular social/behavioral functions (e.g., education, medical care, rehabilitation); and with the exception of most educational environments, they are all what have been characterized as "total institutions."

While these characteristics would seem to suggest such institutions to be important and, in some cases, even ideal settings for environment/behavior research, the literature is more limited than might be expected. Since people are not assigned to such settings in random fashion, the control of traditional laboratory experiments is not easily achieved. Even quasi-experimental field studies are confronted with difficult problems of matching of samples, precisely because people of differing characteristics (e.g., grade level, degree of illness, length of sentence) are sent to very different institutions. Finally, the physical settings which house such social institutions are typically both expensive and long-lived. Experimental manipulations of the environment must often be limited to relatively modest internal reorganization. Only in educational settings, which have experienced a substantial building boom in recent years, and in which competing educational philosophies are often reflected in very different physical settings, have such limitations to environment/behavior research been overcome to any degree.

It should be recognized that the "basic processes" reviewed in Part II are of direct relevance in understanding environment/behavior relationships as they occur in these social institutions. Obvious examples might include the lack of sensory stimulation in a prison environment, the difficulty of way-finding in a large and complex hospital,

or the consequences of over-crowding in a day care center.
Thus the reader is encouraged to take particular note of the
cross-references included at the conclusion of each section.

With respect to each of the several classes of social
settings considered, brief mention might be made of various
key issues raised in the following sections.

a. Educational Environments. Much of the research here
emerges from recent interest in "open education" and the
differences and relationships, between educational program
and physical setting. Other research questions concern the
behavioral consequences of room arrangement and seating
patterns, and evaluation of luminous, thermal, and
especially acoustic features of these environments.

b. Medical Care and Mental Health Environments. These
setting types might be characterized as what Canter and
Canter (item #993) have called "therapeutic" environments.
Within medical care settings, considerable attention has
focused on physical design of the nursing unit and its
behavioral consequences in terms of efficiency, supervision,
and privacy. Much of the research in mental health settings
has focused on patterns of behavior within wards and
dayrooms, often utilizing ecological methodologies. The
recent concern with "normalization" is reflected in both
community-based mental health settings and enrichment of
more traditional psychiatric facilities.

c. Correctional Environments. There is but a scattering
of empirical studies, along with some archival research on
the impacts of group and facility size, available on
correctional environments. Given this paucity of research,
a number of additional citations are included of a
predominantly historical or non-empirical nature.

a. Educational Environments

928. Castaldi, B. CREATIVE PLANNING OF EDUCATIONAL
         FACILITIES. Chicago: Rand McNally & Co., 1969.
         364 pp.

   Presents an insightful, albeit intuitive, approach to
planning and design of educational facilities. Explores the
implications of the psychology of education for physical
design and includes criteria for visual, acoustical and
thermal environments.

929. Coates, G. ALTERNATIVE LEARNING ENVIRONMENTS.
         Stroudsburg, Pa.: Dowden, Hutchinson & Ross,
         1974. 387 pp.

   Reviews much of the work of the past several decades on
alternative approaches to education with a focus on
implications for physical design. Part 3 focuses on
evaluation of educational play environments.

930. King, J., & Marans, R. W. THE PHYSICAL ENVIRONMENT
         AND THE LEARNING PROCESS: A STUDY OF RECENT
         RESEARCH. Ann Arbor, Mi.: Survey Research Center
         and Architectural Research Laboratory, University
         of Michigan, 1979. 85 pp.

   An annotated and indexed bibliography of over 70
entries, with a focus on academic achievement in non-
traditional settings. Additional sections deal with school
size, space and density, color, and the visual and acoustic
environments.

931. Kritchevsky, S., & Prescott, E. PLANNING ENVIRONMENTS
         FOR YOUNG CHILDREN: PHYSICAL SPACE. Washington,
         D. C.: National Association for the Education of
         Young Children, 1969. 56 pp.

Systematic observation of a range of day care centers
indicated availability and placement of materials and
equipment to be among the most effective predictors of
program quality. Reduced quality of outdoor play areas led
to a range of negative consequences.

932. Larson, C. T. THE EFFECT OF WINDOWLESS CLASSROOMS ON
      ELEMENTARY SCHOOL CHILDREN. Ann Arbor, Mi.:
      University of Michigan Press, 1965. 111 pp.

Results of a 3 year study indicated very little pupil
reaction to the presence or absence of windows in their
classroom. Teachers indicated general satisfaction due to
the elimination of outside distractions.

933. Manning, P. (Ed.) THE PRIMARY SCHOOL: AN ENVIRONMENT
      FOR EDUCATION. Liverpool: University of
      Liverpool, Department of Building Science, 1967.
      163 pp.

Provides a variety of general conclusions regarding the
school environment - spatial, visual, thermal, and aural -
based on the observation/assessment of about 20 primary
schools in the northwest of England.

934. Markus, T. A., et al. BUILDING PERFORMANCE. New
      York: Wiley, 1972. 281 pp.

A conceptual model of people/building systems is
developed and utilized as the basis for the appraisal of
performance of school environments. Emphasis is placed upon
the grouping and bounding of spaces within schools and the
implications of such design decisions for building and
student performance.

935. Moore, G. T., Cohen, U. et al. DESIGNING ENVIRONMENTS
      FOR HANDICAPPED CHILDREN: A DESIGN GUIDE AND CASE
      STUDY. New York: Educational Facilities
      Laboratories, 1979. 95 pp.

A behaviorally-based "design guide" for education/play environments for developmentally disabled children. Includes a review of developmental disabilities and the impact of the physical environment on exceptional children as well as a case study based upon the principles of the guide.

936. Moos, R. H. EVALUATING EDUCATIONAL ENVIRONMENTS. San Francisco: Jossey-Bass, 1979. xvi, 334 pp.

Focuses primarily on what is characterized as the social climate, or "personality" of educational environments and on the development of Residence Environment and Classroom Environment Scales   Chapters 4 and 8 deal with organizational, contextual and, to a limited extent, architectural influences in student living and classroom environments respectively.

937. Nimnicht, G., McAfee, O., & Meier, J. THE NEW NURSERY SCHOOL. Morristown, New Jersey: General Learning Press, 1969. 181 pp.

Review of an experiment in early childhood education for disadvantaged students. Part III focuses on the learning environment with a brief discussion of the physical setting of the school.

938. Propst, R. HIGH SCHOOL: THE PROCESS AND THE PLACE. New York: Educational Facilities Laboratories, 1972. 119 pp.

An informal but insightful presentation of the open-plan concept for schools. Includes generic design criteria for spatial organization, visual and auditory privacy, interior surfaces and equipment.

939.   Ambrose, K., & Ambrose, L.   Improving library
       effectiveness through a sociophysical analysis.
       Bulletin of the Medical Libraries Association,
       1977, 65, 430-442.

       Design modifications in a nursing school library, based
upon a sociophysical analysis of the setting, led to an
increase in utilization along with a decline in loss of
library materials. Library users reported more positive
feelings toward both setting and staff.

940.   Artinian, V. A.   The elementary school classroom: A
       study of the built environment.   In #35 (EDRA-2),
       pp. 13-21.

       Findings of the study focus on thermal, luminous, aural
and spatial aspects of the environments, as well as the role
of cultural variables as they relate to general
satisfaction.

941.   Artinian, V. A.   Performance appraisal and the user:
       A paradox.   Industrialization Forum, 1972, 3,
       37-40.

       As the number of students in a classroom increased,
students' satisfaction with available space decreased.
Irrespective of density, teachers tended to express greater
satisfaction with classroom area than did students.

942.   Axelrod, S., Hall, R. V., & Tams, A.   Comparison of
       two common classroom seating arrangements.
       Academic Therapy, 1979, 15, 29-36.

       Classrooms were shifted from table to row arrangement
and back during a one month period in two elementary
schools. Row formation promoted a higher study level and
fewer incidences of talking out behavior.

943. Becker D., & Janzen, H. Behavioral mapping of student
      activity in open-area and traditional schools.
      American Educational Research Journal, 1978, 15,
      507-517.

More locations were utilized and greater interaction
among peers was found to take place in open space schools.
Social behavior, travel and housekeeping activities were
more frequent in open settings, while reading and writing
occurred less often.

944. Becker, F. D., et al. College classroom ecology.
      Sociometry, 1973, 36, 514-525.

Students' grades in a given course were found to
decrease as a function of the distance of their typical seat
from the instructor. There was no significant relationship
between overall grade point average and seating position.

945. Bell, A. E., Switzer, F., & Zipursky, M. A. Open area
      education: An advantage or disadvantage for
      beginners? Perceptual and Motor Skills, 1974, 39,
      407-416.

A study of the achievement of two groups of first
graders, one entering an open-space school and the other a
conventional school.

946. Brody, G., & Zimmerman, B. The effects of modeling
      and classroom organization on the personal space
      of third and fourth grade children. American
      Educational Research Journal 1975, 12, 157-168.

Based upon paper and pencil tests requiring the
positioning of silhouette figures, open classroom students
reported smaller interpersonal distances to specific
individuals than did traditional classroom students.

947.   Brunetti, F. A.   Noise, distraction and privacy in
       conventional and open school environments.   In #36
       (EDRA-3), Vol. 1, Sec. 12.2, pp. 1-6.

       While actual measured noise levels were higher in open
space than in traditional classrooms, open spaces were less
reverberant and were less often reported by students as
being too noisy.

948.   Downing, L. L., & Bothwell, K. H.   Open-space schools:
       Anticipation of peer interaction and development
       of cooperative independence.   Journal of
       Educational Psychology, 1979, 71, 478-484.

       Compared with students from a traditional school, eighth
graders from an open space school selected seats more
conducive to interaction and made significantly more
cooperative responses in a prisoner's dilemma game.
Cooperative behavior was influenced by both sex and race; no
differences in locus of control were found.

949.   Durlak, J. T., Beardsley, B. E., & Murray, J. S.
       Observation of user activity patterns in open and
       traditional plan school environments.   In #36
       (EDRA-3), Vol. 1, Sec. 12.4, pp. 1-8.

       When contrasted with traditional schools, open plan
settings were found to exhibit less structuring of space,
less formal student-teacher contact, and more small group or
individual work.

950.   Dykman, B. M., & Reis, H. T.   Personality correlates
       of classroom seating position.   Journal of
       Educational Psychology 1979, 71, 346-354.

       Students selecting seats on the periphery of classrooms
had less positive self-concepts, less positive attitudes

toward the class, and placed a greater distance between
themselves and a teacher in a projective test than did
pupils sitting in the front of the room.

951.   Evans, G. W., & Lovell, B. Design modifications in an
       open plan school. Journal of Educational
       Psychology, 1979, 71, 41-49.

   The addition of several variable height partitions in an
open-plan school substantially reduced classroom
interruptions. Content questioning by students increased,
while non-substantive, process questioning remained the
same.

952.   Evans, K. The spatial organization of infants'
       schools, Journal of Architectural Research, 1974,
       3(1), 26-33.

   Analysis of spatial configurations in 25 British schools
revealed a wide variety of interior arrangements. Many
vestiges of more traditional approaches to education were
still found, in many cases including centrally located
teachers' desks.

953.   Fagot, B. I. Variations in density: Effect on task
       and social behaviors of pre-school children.
       Developmental Psychology, 1977, 13, 166-167.

   In comparison with their moderate and low density
American counterparts, no negative effects were observed in
very high density Dutch nursery schools. As a result of
their very directed programs, childen in the Dutch schools
spent more time in positive interactions.

954.   Gorton, C. E. The effects of various classroom
       environments on performance of a mental task by
       mentally retarded and normal children. Education
       and Training of the Mentally Retarded, 1972, 7,
       32-38.

Both normal and cultural-familial retarded children
performed simple addition tasks best when visually secluded,
while brain-injured subjects performed best when completely
secluded. There was no significant difference between the
two groups in overall performance.

955. Gottlieb, J., & Budoff, M. Social acceptability of
      retarded children in nongraded schools of
      differing architecture. American Journal of
      Mental Deficiency, 1973, 78, 15-19.

While educable mentally retarded children in an open
school were known more often by their non-retarded peers,
they were selected as friends less frequently and rejected
more often than those retarded children in the walled
school.

956. Gump, P. V. School environments. In #10 (Altman &
      Wohlwill 3) pp. 131-174.

An examination of school environments at all levels from
preschool to university from the perspective of behavioral
ecology and behavior-setting theory. The school is analyzed
in both physical and institutional terms; among the physical
aspects considered are the effects of density and of "open"
classroom design.

957. Gump , P. V., with Good, L. R. Environments operating
      in open space and traditionally designed schools.
      Journal of Architectural Research, 1976 5(1),
      20-27.

An analysis of the relationship between physical setting
and educational program in four schools revealed students in
open plan settings to be spending more non-substantive time
and less time on tasks. Teachers in the open plan settings
were most concerned about noise and lack of program
flexibility.

958. Harper, L. V., & Sanders, K. M. Pre-school children's
     use of space: Sex differences in outdoor play.
     Developmental Psychology, 1975, 11, 119.

     Comparisons of play activity, location and equipment
among nursery school children indicated that girls spent
more time engaged in craft and kitchen activities indoors,
while boys utilized more space as well as sand, tractors and
climbing equipment outdoors.

959. Karmel, L. J. Effects of windowless classroom
     environments on high school students. Perceptual
     and Motor Skills, 1965, 20, 277-278.

     Among a sample of over 1,200 students, it was found that
those attending class in windowless rooms included windows
in drawings of a school significantly more often than those
in windowed classrooms.

960. Konyea, M. Location and interactions in row-and-
     column seating arrangements. Environment and
     Behavior, 1976, 8, 265-282.

     An assessment of the impact of both locational and
personal variables on participation in a traditional
classroom. Location was found to influence behavior of
students previously characterized as "high" or "moderate"
with respect to verbalization rates. As in previous
research, centrally located seats were associated with
higher levels of participation.

961. Krantz, P. J., & Risley, T. R. Behavioral ecology in
     the classroom. In K. D. O'Leary & S. G. O'Leary
     (Eds.), CLASSROOM MANAGEMENT: THE SUCCESSFUL USE
     OF BEHAVIOR MODIFICATION. New York: Pergamon
     Press, 1977.

Provision of separate marked spaces on the floor while listening to teachers' stories resulted in heightened attentiveness on the part of kindergarten children compared to their simply crowding around the teacher. The provision of marked spaces was just as effective as behavioral contingencies in increasing attentiveness.

962.   Lewis, D. T. Noise in primary schools: Implications for design. Journal of Architectural Research, 1977, 6(1), 34-37.

An analysis of classroom activities in relationship to the acoustic properties of different teaching spaces. It is suggested that no single space can satisfy all requirements and that conflicts may be resolved through spatial or temporal separation of activities.

963.   Lukasevich, A., & Gray, R. F. Open space, open education, and pupil performance. The Elementary School Journal, 1978, 79, 108-114.

This study explored the relationship between open and closed instructional style, open and traditional physical environment, and selected cognitive and affective outcomes for 321 third grade pupils in 10 schools. For all but five of the thirteen dependent variables studied, no significant differences were found that were related to program or facility.

964.   Morrison T. L., & Thomas, M. D. Self-esteem and classroom participation. Journal of Educational Research, 1975, 68, 374-377.

Partial support was provided for the hypothesis that students with lower self-esteem prefer to sit further back in the classroom, and contribute less to class interactions.

965. Reiss, S., & Dydhalo, N. Persistence, achievement and
      open space environments. Journal of Educational
      Psychology, 1975, 67, 506-513.

      Second grade students from three open plan schools were
significantly more persistent than students from three
conventional schools. Achievement test scores yielded no
significant differences between the two groups.

966. Rivlin, L., & Rothenberg, M. The use of space in open
      classrooms. In #30 (Proshansky et al., 1976), pp.
      479-489.

      An observational study mapping both occupant behavior
and location of furniture and equipment in eight open
classrooms in two older urban schools. Physical layouts
were found to remain quite stable over the academic year,
and individual work predominated despite the expressed
desire for group activities. The traditional image of the
"front" of the classroom was seen as still shaping teachers'
use of space.

967. Ross, R., & Gump, P. V. Measurement of designed and
      modified openness in elementary school buildings.
      In #41 (EDRA-8), pp. 243-251.

      Observations of 19 open plan schools over several years
indicated an increased utilization of both temporary and
permanent dividers and partitions. Those schools which
initially had been most open were found to have changed the
most. The possibility of an optimum degree of openness is
considered.

968. Sanders, S. G., & Wren, J. P. The open-space school-
      how effective? Elementary School Journal, 1976,
      77, 57-62.

Reviews six studies of the impact of open space schools
on academic achievement and affective reactions. It is
concluded that students in the open space schools did at
least as well as those in traditional schools in measured
cognitive achievement over the three years of the study, and
better during the second and third years. Favorable
attitudes on the part of teachers, parents and pupils were
found to develop in the open schools; concerns about pupil
anxiety in such settings were not supported.

969. Schwebel, A. I., & Cherlin, D. L. Physical and social
      distancing in teacher-pupil relationships.
      Journal of Educational Psdychology, 1972, 63,
      543-550.

Observational and questionnaire data indicate that
students assigned by teachers to the front row in a
classroom are more attentive, and are more positively
perceived, than those in middle and back rows. Comparable
results were obtained when student locations were reassigned
on a random basis.

970. Seidner, C. J. et al. Cognitive and affective
      outcomes for pupils in an open-space elementary
      school: A comparative study. The Elementary
      School Journal, 1978, 78, 208-219.

A comparison of sixth-grade students in open-space and
traditional schools yielded no significant differences in
academic achievement. Attitudinal measures indicated a
stronger preference for school and teacher in the open
setting. Finally self-perceptions in the two schools were
related to internal versus external locus of control.

971. Shure, M. B. Psychological ecology of a nursery
      school. Child Development, 1963, 34, 979-992.

Children's behaviors were classified and counted in five
nursery school sub-settings. Block play areas attracted

more boys than girls, yielded moderately high social
interaction and more destructive behavior than any other
setting. Active social exchange was particularly frequent
in the doll play area and low in the art area.

972. Slater, B.   Effects of noise on pupil performance.
      Journal of Educational Psychology, 1968, 59,
      239-243.

     Standardized reading tasks carried out by junior high
school students for the length of a typical class period
were not affected by the peaks of noise (up to 90 dB)
typical of a normal school environment.

973. Smith, P. Aspects of the playgroup environment.   In
      #47 (Canter & Lee), pp. 56-62.

     Decrease in availability of toys and equipment in a
nursery school setting resulted in increased social
interaction more aggression and stress behaviors, and a
decrease in solitary activity.

974. Smith, P. K., & Connelly, K. Toys, space and
          children. Bulletin of the British Psychological
      Society, 1973, 26, 167.   (Abstract)

     The amount of equipment provided in two children's
playgroups had a more significant effect than did amount of
space on most aspects of social behavior.

975. Sommer, R. Classroom ecology. Journal of Applied
      Behavioral Science, 1967, 3, 489-503.

     Observations indicated a relationship between classroom
participation and seating arrangement in traditional
classroom, seminar and laboratory spaces. Results are seen
as supportive of the expressive contact hypothesis linking
visual contact and interaction.

976.   Sommer, R. The ecology of study areas. Environment
        and Behavior, 1970, 2, 271-280.

    Several thousand students were interviewed while
studying in diverse locations.   Libraries, outdoor spaces,
cafeterias, and place of residence were found to serve
specialized study functions.

977.   Sommer, R., & Olsen, H. The soft classroom.
        Environment and Behavior, 1980, 12, 3-16.

    Conversion of a traditional, straight-row classroom into
a softer space with carpeting, cushions, and other
decorative features resulted in favorable student response
and an increase in classroon participation.

978.   Stebbins, R. A.   Physical context influences on
        behavior: The case of classroom disorderliness.
        Environment and Behavior, 1973, 5, 291-314.

    Several hypotheses are advanced regarding the relation
between physical conditions in classrooms and disorderly
behavior. Issues raised pertain to open design, lack of
privacy, and layout. (Hypotheses were not tested.)

979.   Stires, L. Classroom seating location, student grades
        and attitudes: Environment or self-selection.
        Environment and Behavior, 1980, 12, 241-254.

    Irrespective of free self-selection or alphabetical
assignment of classroom seat, students in the center of the
room received higher grades and liked both course and
instructor more than did students seated at the sides.

980. Traub, R. E., et al. Closure on openness: Describing
      and quantifying open education. Interchange,
      1974, 3, 69-84.

      Questionnaire data were gathered from 449 teachers in 30
schools of varying architectural design. While teachers
from physically open schools reported that their programs
were open more often than did teachers from traditional
schools, the differences were not large and the distribution
of scores for open and traditional buildings overlapped
considerably.

981. Walberg, H. Physical and psychological distance in
      the classroom. The School Review, 1969, 77,
      64-70.

      A self-report study relating attitude toward school and
self with seat preference. Students preferring the front of
the room placed a high positive value on learning while
those at the rear and sides took a more negative attitude.

982. Walsh, D. P. Noise levels in open plan educational
      facilities. Journal of Architectural Research,
      1975, 4(2), 5-16.

      Measurement of background noise levels were made in a
number of open plan schools. Data suggested a direct
relationship between numbers of students and surface
finishes in a space and noise levels. Spatially sub-divided
teaching areas resulted in greater fluctuations in noise
level, and higher levels of annoyance, than fully open
spaces.

983. Weinstein, C. S. Modifying student behavior in an
      open classroom through changes in physical design.
      American Educational Research Journal, 1977, 14,
      249-262.

Relatively minor physical design changes within an
elementary school classroom were found to influence the
spatial distribution of students within the room, the
frequencies of specific categories of behavior, and the
range of behaviors within particular locations.

984. Weinstein, C. S. The physical environment of the
        school: A review of the research. Review of
        Educational Research, 1979, 49, 577-610.

A thorough and thoughtful review of research on the
behavioral and attitudinal consequences of the classroom
environment. Studies of specific environmental variables
(e.g., seating position, noise, density) are reviewed, along
with the more limited research conducted from an
"ecological" perspective, and studies of the effects of open
space school designs.

985. Weinstein, C. S., & Weinstein, N. D. Noise and
        reading performance in an open space school.
        Journal of Educational Research, 1979, 72,
        210-213.

Quiet versus normal [47 dB(A) and 60 dB(A)] background
noise levels in a classroom did not result in a differential
effect on the reading performance of fourth grade pupils.

986. Wright, R. J. The affective and cognitive
        consequences of an open education elementary
        school. American Educational Research Journal,
        1975, 12, 449-465.

Various pupil outcome dimensions were compared for an
open and a traditional elementary school. After two and one
half years, no differences were found for three measures of
personality or three measures of cognition. Students in the
open school did have significantly lower levels of academic
achievement, and direct observations revealed the expected
differences in teaching approach in the two settings.

987. Zifferblatt, S. M. Architecture and human behavior:
      Toward an understanding of a functional
      relationship. Educational Technology, 1972, 12,
      54-57.

The attention span of elementary school students was
found to be shorter in a classroom where relatively large
numbers of student desks were clustered together and areas
for different activities were not clearly designated and
separated.

988. Chambers, J. A. A study of attitudes and feelings
      toward windowless classrooms. Dissertation
      Abstracts International, 1964, 24, 4498.

989. Gingold, W. The effects of physical environment on
      children's behavior. Dissertation Abstracts
      International, 1971, 32-A, 2947.

990. Johnson, R. H. The effects of four modified elements
      of a classroom's physical environment on the
      socio-psychological environment of a class.
      Dissertation Abstracts International, 1973, 34-A,
      1002.

991. Wang, Y. T. The result of differential seating
      arrangements upon students' anxiety level,
      acquaintance volume, and perceived social
      distance. Dissertation Abstracts International,
      1973, 33-A, 4191.

992. Warner, J. B. A comparison of students' and teachers'
      performance in an open area facility and in self-
      contained classrooms. Dissertation Abstracts
      International, 1971, 31-A, 3851.

See also: citations 331, 462, 574, 577, 585, 596, 598 608.

b. Medical Care and Mental Health Environments

993. Canter, D. & Canter S. DESIGNING FOR THERAPEUTIC
        ENVIRONMENTS. New York: Wiley, 1979. 352 pp.

    A review of research dealing with a range of
institutions which take account of the physical environment
in therapeutic processes. Includes evaluations of settings
for children, the developmentally disabled and aged, as well
as acute general and psychiatric hospitals.

994. Cleveland State Hospital, Research Department. A
        BIBLIOGRAPHY OF MATERIALS USEFUL FOR CHANGE IN
        MENTAL HOSPITALS: ARCHITECTURE, INSTITUTIONAL
        SETTINGS AND HEALTH. Monticello, Ill,: Council
        of Planning Librarians, 1973. (Exchange
        Bibliographies, Nos. 463, 464, & 465) 464 pp.

    A very extensive, non-annotated bibliography on mental
health environments, organized in 49 topical catagories with
an emphasis on psycho-social issues.

995. Cleveland State Hospital, Research Department. A
        BIBLIOGRAPHY OF MATERIALS USEFUL FOR CHANGE IN
        MENTAL HOSPITALS: ARCHITECTURE, INSTITUTIONAL
        SETTINGS, AND HEALTH. Monticello, Ill,: 1974.
        (Exchange Bibliographies, Nos. 533 & 534) 80 pp.

    A supplement to Exchange Bibliographies 463-465 (see
preceding item), organized by the same categories and non-
annotated.

996. Cohen, U., & Pencikowski, B. User/environment
        interface in health care systems. Monticello,
        Ill.: Council of Planning Libraries, 1973.
        (Exchange Bibliographies, No. 367) 27 pp.

    A non-annotated bibliography organized into ten sections
including materials on social/psychological factors in
health care environments, human factors, and issues of
spatial design and physical organization.

997. Good, L. R., Siegel, S. M. & Bay, A. P. (Eds.)
        THERAPY BY DESIGN: IMPLICATIONS OF ARCHITECTURE
        FOR HUMAN BEHAVIOR. Springfield, Ill.: C. C.
        Thomas, 1965. 193 pp.

    Includes the presentations and dialogue of an early
conference focusing on the role of the physical environment
in psychiatric therapy. The history of the classic
Kirkbride hospital is reviewed along with proposals for the
behaviorally-based renovation of one such building.

998. Knight, R. D., Weitzer, W., & Zimring, C. OPPORTUNITY
        FOR CONTROL AND THE BUILT ENVIRONMENT: THE
        EFFECTS OF THE LIVING ENVIRONMENT ON THE MENTALLY
        RETARDED. Amherst, Mass.: The Environmental
        Institute, University of Massachusetts, 1978. 308
        pp.

    A four year longitudinal study of interior design
changes at a large state mental hospital demonstrated that
physically "normalized" environments had positive impacts on
the behavior patterns of residents. Emphasis is placed upon
the degree to which settings offer occupants opportunities
for control of their environment.

999. Lindheim, R., Glaser, H., & Coffin, C. CHANGING
        HOSPITAL ENVIRONMENTS FOR CHILDREN. Cambridge,
        Mass.: Harvard University Press, 1972. xii, 206
        pp.

Based upon the available, albeit limited literature, the
environmental needs of infants, toddlers, school-age, and
adolescent children are discussed in relationship to
specific developmental needs. On this basis, extensive
design guidelines for children's hospitals are presented.

1000.   Sears, D., & Auld, R. HUMAN VALUATION OF COMPLEX
         ENVIRONMENTS. London: Joint Unit for Planning
         Research, University College & London School of
         Economics, 1976.

Surveys of over 400 patients in 16 hospitals indicated
surveillance and patients' sense of security, provision of
privacy within sanitary facilities, and spaces for patients
out of bed to be the most important factors in evaluation of
the wards in which they were housed.

1001.   Thompson, J. D., & Goldin, G. THE HOSPITAL: A
         SOCIAL AND ARCHITECTURAL HISTORY. New Haven:
         Yale University Press, 1975. xxviii, 349 pp.

The history of hospital design is reviewed in terms of
four functional determinants: a healthful environment;
privacy; efficiency of layout; and adequate supervision.
The final third of the book discusses the "Yale studies,"
which focus on patient preferences for hospital furnishings,
the utilization of call systems, and the impact of nursing
unit layout on staff efficiency.

1002.   Anderson, H. E., Good, L. R., & Hurtig, W. E.
         Designing a mental health center to replace a
         county hospital. Hospital and Community
         Psychiatry, 1978, 27, 807-813.

Reviews the planning, programming and design of a new
active-treatment facility, as well as the results of an

informal evaluation conducted one year after completion of
the center.

1003. Angell, M. F., & Spaeth, G. L. Multibed rooms
         improve morale, patient survey shows. Hospitals,
         1968, 42, 57-58.

    Data gathered from 510 patients at admission to and/or
discharge from an ophthalmological hospital indicated a
strong preference for multi-bed rooms. Comparisons by sex,
age, degree of vision or cost of accomodation showed no
significant differences.

1004. Becker, F. D., & Poe, D. B. The effects of user-
         generated design modifications in a general
         hospital. Journal of Nonverbal Behavior, 1980, 4,
         195-218.

    Relatively minor changes in one wing of a small hospital
resulted in increased use of public spaces, and changes in
behavior of various user groups and in organizational
climate. Mood and morale and perceived quality of health
care increased for patients and staff but decreased for
visitors.

1005. Canter, D. Children in hospital: A facet theory
         approach to person/place synomorphy. Journal of
         Architectural Research, 1977, 6, 20-32.

    Guttman's "facet theory" is utilized to explore the
relationships between patterns of activity on and experience
of a children's ward, and aspects of the physical
organization of the ward. Analysis reveals an isolation of
non-medical activities quite different from the open and
integrated ward intended by the designers.

1006. Canter, D. Royal Hospital for Sick Children: A
         psychological analysis. Architects Journal, 1972,
         156, 525-564.

Different user groups' evaluations of a large district
hospital were clearly related to their varying goals.
Analysis of the process of planning and design of the
facility revealed the impact of constraints and
preconceptions under which architects and staff operated.

1007. Cheek, F. E., Maxwell, R., & Weisman, R. Carpeting
       the ward: An exploratory study in environmental
       psychology. Mental Hygiene, 1971, 55, 109-118.

Installation of carpeting in two wards of a state
institution was viewed positively by patients and
administrators, but with mixed feelings by staff. Anecdotal
reports suggest an improvement in personal care and self-
maintenance behaviors of patients.

1008. Commock, R. Confidentiality in health centers and
       group practices: Use implications for design.
       Journal of Architectural Research, 1975, 4(1),
       5-17.

Discussion of issues of privacy and confidentiality of
patient information based upon an analysis of 32 British
health centers.

1009. Davis, G. A. Patient attitudes related to some
       problems in psychiatric architecture.
       Psychological Reports, 1963, 12, 467-478.

A review of interviews with 32 open ward patients
regarding such issues as preferred ward size, dining and
sleeping arrangements, and centralized versus dispersed
buildings. Article serves to highlight significant changes
in mental hospital design in the past two decades.

1010. Fairbanks, L., et al. The ethological study of four
       psychiatric wards: Patient, staff and system
       behaviors. Journal of Psychiatric Research, 1977,
       13, 193-209.

     Behavioral observations were carried out in two
university and two veterans hospital wards. Data were
gathered on behavioral profiles of patient and staff, the
social organization of each ward, and the impact of social
and physical environments upon behavior.

1011. Holahan, C. J. Seating patterns and patient behavior
       in an experimental day room. Journal of Abnormal
       Psychology, 1972, 80, 115-124.

     Psychiatric patients were found to exhibit a greater
amount of social interaction when day room furniture was
arranged in a fully or partially sociopetal configuration as
opposed to a sociofugal or unstructured configuration.
Nonsocial activity, however, was not influenced by seating
arrangement. Finally, patients evidenced a preference for
the sociopetal seating.

1012. Holahan, C. J. Environmental change is a psychiatric
       setting: A social systems analysis. Human
       Relations, 1976, 29, 153-166.

     The extensive physical remodeling of an admissions ward
precipitated changes in the ward social system in the areas
of staff role relationships, distribution of power, and
communication styles. These changes mediated the effects of
physical change on patient behavior. A model of a time-
ordered process of environmental change is presented.

1013. Holahan, C. J., & Saegert, S. Behavioral and
       attitudinal effects of large-scale variation in
       the physical environment of psychiatric wards.
       Journal of Abnormal Psychology, 1973, 82, 454-462.

Patients on an extensively remodeled admissions ward
showed significantly more socializing and less passivity
than on a non-remodeled control ward. The most active
patients on the experimental ward expressed a positive
attitude toward their physical environment while the most
active patients on the control ward expressed negative
attitudes.

1014.   Hunter, A., Schooler, C., & Spohn, H. The
        measurement of characteristic patterns of ward
        behavior in chronic schizophrenics. Journal of
        Consulting Psychology, 1962, 26, 69-73.

Reviews the development of a Location-Activity-Inventory
and its utilization in two closed wards over a six month
period. Results reveal very limited social interaction and
a high degree of immobility.

1015.   Kenny, C., & Canter, D. Evaluating acute general
        hospitals. In #993 (Canter & Canter), pp.
        309-332.

A review of research on the relationship between the
physical settings of hospitals and the goals of
practitioners and patients. The hospital as a whole is
considered first, and then the ward. It is suggested that a
"technological fix" approach has led to overly mechanistic
and inhumane hospital design.

1016.   Kreger, K. C. Compensatory environment programming
        for the severely retarded behaviorally disturbed.
        Mental Retardation, 1971, 9, 29-32.

Description of a program in a state hospital designed to
reduce environmental stress through reduction of
overcrowding and an increase in sensory stimulation and
variation in pattern of daily life. Informal evaluation of
the program indicates a marked diminution in behavioral
problems, and increased behavioral adjustment and
acquisition of skills.

1017.   Ittelson, W. H., Proshansky, H. M., & Rivlin, L. G.
        Bedroom size and social interaction of the
        psychiatric ward. Environment and Behavior, 1970,
        2, 255-270.

   Observations were made of social, isolated active, and
isolated passive behaviors on four psychiatric wards in
three hospitals, with bedrooms accommodating from one to
twelve patients. Isolated passive behavior was the most
common type of activity in all bedrooms, but its proportion
of total activity increased with the size of the room;
social activity decreased with the number of patients
assigned to the room.

1018.   Lawton, M. P., Liebowitz, R., & Charon, H. Physical
        structure and the behavior of senile patients
        following ward remodeling. Aging and Human
        Development, 1970, 1, 231-239.

   The change from shared four-bed rooms to a cluster of 6
private rooms with shared social space halved time spent by
patients in their rooms and almost trebled locomotion into
circulation spaces.

1019.   LeCompte, W. F. The taxonomy of a treatment
        environment. Archives of Physical Medicine and
        Rehabilitation, 1972, 53, 109-114.

   Results of a twelve month behavior setting survey in a
rehabilitation institute indicate that patients inhabit only
a small segment of the treatment environment where their
primary exposure is to staff with little formal professional
training. Physicians were found to have control over the
largest proportion of the 122 settings in the hospital.

1020.   Levy, E., & McLeod, W. The effects of environmental
        design on adolescents in an institution. Mental
        Retardation, 1977, 15, 28-32.

The day room of a state hospital was "enriched" through the creation of differing areas and the introduction of play apparatus and modular learning booths. There was a significant increase in five socially acceptable behaviors and a significant decrease in neutral activity.

1021.   Lippert, S. Travel in nursing units. Human
        Factors, 1971, 13, 269-282.

Description of a model for determination of nurses' travel time on the basis of both physical layout and typical patterns of work. The model is utilized to compare a number of different nursing unit plans.

1022.   McClannahan, L. E., & Risley, T. R. Design of living
        environments for nursing home residents:
        Increasing participation in recreation activities.
        Journal of Applied Behavior Analysis, 1975, 8,
        261-268.

Provision of manipulative materials in the lounge area of a skilled-care nursing home did not increase the number of residents using the area, but did result in a trebling of activity levels when staff prompts were provided.

1023.   McLaughlin, H. Evaluating the effectiveness of
        innovative design for a community mental health
        center. Hospital and Community Psychiatry, 1976,
        27, 566-571.

Describes the design of a community mental health center as well as the results of two informal evaluations, one conducted shortly after completion and the second five years later.

1024. Nehrke, M. F., et al. Health status, room size, and
activity level: Research in an institutional
setting. Environment and Behavior, 1979, 11,
451-463.

Level of activity of residents in a Veterans
Administration domiciliary was found to decrease as number
of occupants per room increased; contrary to this pattern,
however, residents of the largest room were the second most
active group. Possible explanations for this anomalous
finding are considered, based upon the impact of age
differences as well the possibility of a curvilinear
relationship between room size and activity level.

1025. Pelletier, R., & Thompson, J. D. Yale Index measures
design efficiency. Modern Hospital, 1960, 95 (5),
73-77.

Details development of the Yale Traffic Index for
evaluation of efficiency of alternative nursing units.
Analyses of 19 floor plan configurations of nursing units
suggested that efficiency is not directly related to unit
size and, within the range of sizes studied, layout is the
most important factor.

1026. Piedmont, E., & Dornblaser, B. Evaluation of patient
welfare. Health Services Research, 1970, 5,
248-257.

Work sampling and direct time studies were utilized to
measure patient-staff contact in both conventional and
radial hospital nursing units.

1027. Pill, R. Space and social structure in two
children's wards. Sociological Review, 1967, 151,
179-192.

A sociological analysis of two wards differing in physical design suggests that the impact of settings is reflected in the way they are used by the staff in acting their roles rather than through any direct influence upon behavior.

1028. Plutchik, R., Schulman, R. & Conte, H. Territorial and communication patterns of patients and staff members in a mental hospital. Mental Hygiene, 1972, 56, 102-104.

Random observations of two state hospital wards over a one month period demonstrated that very little interaction occurred among patients or between patients and staff. For staff, the nurses' office was the most frequented location while patients were found either in the day room or the bedrooms.

1029. Rago, W. V., Parker, R. M., & Cleland, C. C. Effect of increased space on the social behavior of institutionalized profoundly retarded male adults. American Journal of Mental Deficiency, 1978, 82, 554-558.

A 29% increase in the size of a day room occupied by 14 residents was found to significantly reduce the incidence of aggressive acts occurring within the group. It is suggested that the increased space may have reduced aggression by reducing the frequency of inter-personal interactions.

1030. Reizenstein, J. E., & McBride, W. A. Design for normalization: A social environmental evaluation of a community for mentally retarded adults. Journal of Architectural Research, 1977, 6(1), 10-23.

A multi-method evaluation of the living units within this community and their relationship to three issues in normalization: social contact, activity support, and symbolic identification.

1031. Rivlin, L. G., & Wolfe, M. The early history of a
       psychiatric hospital for children. Environment
       and Behavior, 1972, 4, 33-72.

Beginning with initial administrative and planning
decisions, this longitudinal study focuses on: (1) the
relationship between intentions of designer and
administration and actual use of facility: (2) physical and
functional changes made by occupants; and (3) patterns of
utilization of various spaces over the first half year of
occupancy.

1032. Ronco, P. G., Human factors applied to hospital
       patient care. Human Factors, 1972, 14, 461-470.

The experience of hospitalization is characterized in
terms of physical and psychological confinement, lack of
privacy, lack of familiar support, and disruptions of
familiar behavior patterns, all contributing to a loss of
personal control and an increase in stress.

1033. Rosengren, W. R. & DeVault, S. The sociology of time
       and space in an obstetrical hospital. In E.
       Friedson (Ed.), THE HOSPITAL IN MODERN SOCIETY.
       New York: The Free Press of Glencoe, 1963. Pp.
       266-292.

Observations over a four-month period revealed the
importance of both the time and locus of interactions among
staff and patients. Physical segregation of areas served to
indicate gradations in status, while symbolic forms of
segregation appeared to articulate roles.

1034. Sommer, R., & Kroll, B. Mental patients and nurses
       rate habitability. In #993 (Canter & Canter), pp.
       199-212.

Assessment of the habitability of two large state hospitals for the mentally ill and retarded revealed quality of staff and degree of privacy in bedroom areas to be most important to residents. Staff were particularly concerned with relationships between staffing patterns, surveillance and privacy.

1035. Sommer, R., & Dewar, R. The physical environment of
       the ward. In E. Friedson (Ed.), THE HOSPITAL IN
       MODERN SOCIETY. New York: The Free Press of
       Glencoe, 1963. Pp. 319-342.

A variety of key behavioral issues in ward design are presented. Several research studies in geriatric wards within a psychiatric hospital are also reviewed; these focus on the impact of seating arrangements on social interaction, the nature of those patients who spent their time in corridors, and the relationship between ward satisfaction and length of stay in the hospital.

1036. Srivastava, R. K. Undermanning theory in the context
       of mental health care environments. In #38
       (EDRA-5), Vol. 8, pp. 245-258.

Analyzes the behavior and functioning of patients in a state mental hospital in terms of an interaction of undermanning with regard to staff and overmanning in regard to number of patients.

1037. Stembridge, D. A., & LeCompte, W. F. Attitude
       changes after brief encounters with mental
       patients in residential and institutional
       settings. In #36 (EDRA-3), Vol. 1, Sec. 12.7, pp.
       1-6.

Encountering patients within the context of a community setting contributed to significantly more benevolent and less socially restrictive attitudes toward mental patients and mental illness on the part of university undergraduates.

A similar encounter with the same patients in a hospital
setting had no apparent impact upon the students' attitudes.

1038. Thompson, J. D., & Pelletier, R. J. Privacy versus
        efficiency in the inpatient unit. Hospitals,
        1962, 36, 53-57.

    The development of a Privacy Index for nursing units and
its use in conjunction with the Yale Traffic Index is
described. Thirty inpatient units were then evaluated in
terms of both privacy and efficiency. Efficiency was found
to be unrelated to privacy but strongly related to unit
layout.

1039. Trites, D. K. et al. Influence of nursing-unit
        design on the activities and subjective feelings
        of nursing personnel. Environment and Behavior,
        1970, 2, 303-334.

    Radial, double-corridor, and single-corridor nursing
units, all located within the same hospital, were compared.
In general, the radial unit was found most desirable and the
single-corridor unit least desirable. Nursing personnel on
radial units spent significantly less time in travel, were
absent less often and preferred these units.

1040. Willems, E. P. The interface of the hospital
        environment and patient behavior. Archives of
        Physical Medicine, 1972, 53, 115-122.

    Based upon an observational study of 12 rehabilitation
patients, data are reported on frequency and distribution of
patient behavior by location and type, involvement of
hospital personnel in behavior of patients, and the impact
of settings upon independence and "zest" of patient
behavior

1041.   Willems, E. P. Behavioral ecology, health status,
        and health care: Applications to the
        rehabilitation setting. In #8 (Altman & Wohlwill
        1), pp. 211-263.

    Reviews a variety of findings from .long term study of
the rehabilitation process of spinal cord injury victims,
both during and after hospitalization. Behavioral ecology
is presented as a model which can guide and influence
research in health care environments and other institutional
settings.

1042.   Willer, B. et al. Activity patterns and the use of
        space by patients and staff on the psychiatric
        ward. Canadian Psychiatric Association Journal,
        1974, 19, 457-482.

    An observational study of activity type and location in
three wards of a large metropolitan mental hospital.
Patients spent a sizable proportion of their time in
isolated passive behaviors, primarily in their bedrooms, and
staff spent little time interacting with patients. Mention
is made of the remarkable day-to-day consistency of location
and activity of patients.

1043.   Wilson, L. Intensive care delirium: The effect of
        outside deprivation in a windowless unit.
        Archives of Internal Medicine, 1972, 130, 225-226.

    Clinical records were reviewed for 50 patients in each
of two intensive care units, one with views to the outside
and the other without windows. Over twice as many episodes
of organic delirium occurred in the windowless unit. It is
concluded that the lack of windows contributes to sensory
deprivation.

1044.   Wolfe, M. Room size, group size, and density:
        Behavior patterns in a children's psychiatric
        facility. Environment and Behavior, 1975, 7,
        199-224.

Observational data gathered over a 2 1/2 year period
indicate that the "potential density" of patient bedrooms
(ie. square feet per person based on capacity occupancy) was
not related to bedroom use or behavior patterns. Potential
density appears as an inappropriate basis for allotment of
space in the programming of such facilities.

1045. James, E. V. Environment and behavior of hospital
       patients in psychiatric and medical wards.
       Dissertation Abstracts International, 1973, 33-B,
       5494.

See Also: citations 496,516,599.

c. Correctional Environments

1046. Barback, B. Understanding the prison environment: A
       selected interdisciplinary bibliography.
       Monticello, Ill.: Council of Planning Librarians,
       1974. (Exchange Bibliographies, No. 632.) 13 pp.

A non-annotated and wide ranging bibliography of over
280 references, dealing with criminology and design in
general.

1047. Dickens, P., McConville, S., & Fairweather, L. PENAL
       POLICY AND PRISON ARCHITECTURE. London: Barry
       Rose Publishers, 1978, 96 pp.

A set of five essays based upon a symposium at the
University of Surrey in 1977; provides considerable useful
background information, albeit little in the way of
environment-behavior research.

1048. DiGennaro, G. (Ed.), PRISON ARCHITECTURE: AN
       INTERNATIONAL SURVEY OF REPRESENTATIVE CLOSED
       INSTITUTIONS AND ANALYSIS OF CURRENT TRENDS IN
       PRISON DESIGN. London: The Architectural Press,
       Ltd., 1975. 239 pp.

Based upon a research project of the United Nations
Social Defense Research Institute, this volume includes
documentation on 27 prisons in 11 countries, several
articles on prison programming and design, an historical
review, and an annotated bibliography.

1049. Hyde, D. M., & Conway, D. Stress factors as
       identified by research in prisons. Monticello,
       Ill.: Council of Planning Librarians, 1977.
       (Exchange Bibliographies, No. 1244). 19 pp.

Includes 65 citations plus an additional nine
annotations of unpublished research, but with a relatively
limited environment/behavior focus.

1050. Johnston, N. THE HUMAN CAGE: A BRIEF HISTORY OF
       PRISON ARCHITECTURE. New York: Walker and
       Company, 1973. 68 pp.

A primarily historical rather than psychological
analysis of the development of prison architecture. It is
argued that even modern prisons deprive their occupants of
privacy, dignity, and self-esteem.

1051. Moos, R. H. EVALUATING CORRECTIONAL AND COMMUNITY
       SETTINGS. New York: Wiley, 1975. xii, 377 pp.

While   the primary focus of the book is the development
of scales for the assessment of "social climate," some
limited material is included on architectural variables as
well as the impact of facility size and staffing.

1052. Moyer, F. D. et al. GUIDELINES FOR THE PLANNING AND
         DESIGN OF REGIONAL AND COMMUNITY CORRECTIONAL
         CENTERS FOR ADULTS. Urbana, Illinois: Department
         of Architecture, University of Illinois, 1971.
         (loose-leaf, unpaginated)

An extensive set of guidelines for the identification of
correctional problems and development of alternative
classification, routing and treatment programs, as well as
for actual facility planning and design.

1053. Nagel, W. G. THE NEW RED BARN: A CRITICAL LOOK AT
         THE MODERN AMERICAN PRISON. New York: Walker &
         Co., 1973   196 pp.

A study based upon on-site observation and evaluation of
over 100 correctional institutions, involving contact with
both inmates and staff and analysis of facility plans.
Current trends in both design and treatment are reviewed,
and new directions for the criminal justice system and
institutional design are proposed.

1054. Sommer, R. TIGHT SPACES: HARD ARCHITECTURE AND HOW
         TO HUMANIZE IT. Englewood Cliffs, N. J.:
         Prentice-Hall, 1974. 150 pp.

Prisons are presented as the paradigm of "hard
architecture," monumental, fortress-like and regimented in
character, increasingly typical of our social institutions.
Chapters 2 and 3 focus specifically on prison environments,
alternative models of incarceration, and the behavioral
costs of imprisonment.

1055.   Farbstein, J. A.   Juvenile services center program.
        In W. F. E. Preiser (Ed.), FACILITY PROGRAMMING.
        Stroudsburg, Pa.: Dowden, Hutchinson & Ross,
        1978. Pp. 67-84.

        A behaviorally-based architectural programming process
focusing on requirements for space, circulation, ambient
conditions, safety and security, furnishings, flexibility
and site development.

1056.   Fleising, U. The social and spatial dynamics of a
        prison tier: An exploratory study. Man-
        Environment Systems, 1973, 3, 187-195.

        An observational study of one tier in a minimum security
youth correctional center. The data suggest that at least
one group of inmates sharing similar psychiatric diagnoses
displayed similar behavioral and spatial patterning. It is
further proposed that the presence of individual rooms
served to decrease the aggressive potential of the tier.

1057.   Flynn, E. E., & Moyer, F. D. Corrections and
        architecture: A synthesis. Prison Journal, 1971,
        51, 43-53.

        A discussion of the guidelines for correctional planning
developed by the National Clearinghouse for Criminal Justice
Planning and Architecture (item #1052).

1058.   Gill, H. B. Correctional philosophy and
        architecture. Journal of Criminal Law,
        Criminology and Police Science, 1962, 53, 312-322.

        Relations between correctional philosophy and the built
environment are reviewed and four physically distinct
institutional settings are proposed. The variable of size
of group living unit is emphasized.

1059. Glazer, D. Architectural factors in isolation
promotion in prisons. In #29 (Proshansky et al.,
1970), pp. 455-463.

An analysis of the impact of private cells and crowded
and uncrowded dormitories upon prisoners' activities and the
process of rehabilitation.

1060. Hahn, R. R. Behavioral evaluation of a juvenile
treatment center: Case study of a planning
methodology. In #37 (EDRA-4), Vol. 1. pp.
138-149.

Occurrence of self-regulating and infringing behaviors
was only partially correlated with staff expectations
regarding the location where these behaviors were most
likely to occur. Prevalence of these behaviors varied
greatly from setting to setting within the center; physical,
sociological and administrative reasons for these
differences are considered.

1061. Hickey, W. Modern correctional design. Crime and
Delinquency Literature, 1972, 5, 116-121.

A broad review of innovations in jail, prison, and
juvenile institution programming and design, as well as 70+
annotations of articles drawn primarily from architectural
and correctional journals.

1062. Jessness, C. Comparative effectiveness of two
institutional treatment programs for delinquents.
Child Care Quarterly, 1972, 1, 119-130.

A comparison of residents in a 20-bed cottage versus one
of 50 beds at a juvenile correctional ranch. Parole
violations were substantially lower for residents of the
smaller cottage. It is noted that the two settings differed
in terms of staff and treatment as well as physical setting.

1063.   Megargee, E. I. The association of population
        density, reduced space and uncomfortable
        temperatures with misconduct in a prison
        community. American Journal of Community
        Psychology, 1977. 5, 289-298.

    Data gathered at a medium security prison indicated a
significant negative correlation between density and the
rate and number of misconduct incidents.

1064.   Nacci, P. L., Teitelbaum, H. E., & Prather, J.
        Population density and inmate misconduct rates in
        the federal prison system. Federal Probation,
        1978, 41, 26-31.

    Data on inmate misconduct and population density levels
were gathered from 37 federal correctional institutions for
a three year period. Positive and significant correlations
were found between density and three types of incidents.
The relationship between density and misconduct,
specifically with respect to assaults on other inmates, was
highest in institutions housing young adults.

1065.   Paulus P., McCain, G., & Cox, V. Death rates,
        psychiatric commitments, blood pressure and
        perceived crowding as a function of institutional
        crowding. Environmental Psychology and Nonverbal
        Behavior, 1978, 3, 107-116.

    Inmate blood pressure was found to be higher in more
crowded housing, with the degree of perceived crowding more
strongly related to space per person than number of inmates
per dwelling unit. Analysis of archival data indicated
higher rates of death and psychiatric commitment in years of
higher prison population.

1066. Prestholdt, P. H., Taylor, R. R., & Shannon, W. T.
       The correctional environment and human behavior:
       A comparative study of two prisons for women. In
       #40 (EDRA-7), pp. 145-149.

       Staff and inmates were relocated from an old and crowded
"camp-like" setting to a sophisticated new prison. Inmates'
private rooms were more positively evaluated than the open
dorms of the old setting. The size, layout and
sophisticated surveillance equipment of the new prison,
however, led to social contact among smaller and more
diffused groups of inmates as well as reduced inmate-staff
contact.

1067. Schwartz, B. Deprivation of privacy as a functional
       pre-requisite: The case of the prison. Journal
       of Criminal Law, Criminology and Political
       Science, 1972, 63, 229-239.

       It is argued that the suspension of the right of inmates
to privacy is not a consequence of the legal status of
prisoners, but rather of the nature of prison as a social
institution.

1068. Sims, W. R. The half-way house: A diagnostic and
       prescriptive evaluation. In #40 (EDRA-7), pp.
       150-156.

       A qualitative evaluation of three half-way houses for
released convicts. Site inspection and a limited number of
in-depth interviews yielded recommendations regarding
privacy, facility size, recreation, supervision, lighting,
appearance, and location.

1069. Smith, D. E., & Swanson, R. M. Architectural reform
       and corrections: An attributional analysis.
       Criminal Justice and Behavior, 1979, 6, 275-292.

An analysis of the extent to which positive and negative
feelings were attributed to the environment, others and
oneself before and after the move from an antiquated to a
modern correctional facility. For both inmates and staff,
attributions to the environment became more positive after
the move; attributions to cohorts (e.g., other inmates or
staff), however, became more negative.

1070.   Smith, D. E., & Swanson, R. M.   Privacy and
        corrections: A social learning approach.
        Criminal Justice and Behavior, 1979, 6, 339-357.

        Inmate values and expectations with respect to privacy
were assessed in six correctional institutions. Within-
institution differences in living arrangements (e.g., number
of occupants per cell) affected values for privacy, while
differences in expectations were found across institutions.
It is suggested that institutional differences are due more
to variations in the overall institutional environment than
to variations in living conditions.

1071.   Sommer, R.   The social psychology of the cell
        environment.   In NEW ENVIRONMENTS FOR THE
        INCARCERATED.   U. S. Department of Justice, Law
        Enforcement Administration, National Institute of
        Law Enforcement and Criminal Justice, Washington,
        D. C., 1972.   Pp. 15-21.

        A consideration of the possible behavioral consequences
of lack of privacy, stimulus deprivation and crowding as
experienced by inmates of correctional institutions.

1072.   Stokols, D., & Marrero, D. G.   The effects of an
        environmental intervention on racial polarization
        in a youth training school.   In #40 (EDRA-7), pp.
        125-137.

        A quasi-experimental study of the impact of a sociopetal
furniture arrangement within a dayroom upon level of

resident aggression, racial polarization, ratings of
satisfaction of fellow inmates, evaluation of staff, and
positive inter-racial contact. Predicted changes occurred
for only the last two of these variables. The authors
speculate on the likely limited impact of such a specific
physical intervention.

1073. Sykes, G. The prisoner's status as conveyed by the
          environment. In #29 (Proshansky et al, 1970), pp.
          453-455.

    A consideration of the deprivations, both sensory and
symbolic, which are a part of the prison environment.

1074. Van der Ryn, S. Can architecture aid a therapeutic
          process? American Journal of Corrections, 1969,
          31, 41-44.

    Basic environment-behavior concepts are applied to the
planning of small scale Model Treatment Units.

See also: citations 442, 449.

11. THE HOUSING ENVIRONMENT

The relationship between people and the settings in which they live represents a theme running through much environment-behavior research. Thus the housing environment has been studied from a variety of theoretical and methodological perspectives and at differing scales.

The study of people's satisfaction with their housing, particularly as this relates to patterns of residential choice and mobility, has in large part been undertaken by geographers, sociologists, planners and family economists. Variables considered are often aggregate in nature, focusing on proximity to desired locations (e.g., place of work) or people (e.g., those of similar or preferred socio-economic background). More fine-grained features of the dwelling unit itself and its immediate neighborhood context have also been incorporated in some of the recent research in this area.

Much of the work of psychological and architectural researchers has dealt with groupings of dwelling units as these occur in high and low-rise housing and clusters of buildings. Beginning with the classic study of Festinger et al. (item #1080) the relationship between spatial layout and patterns of social interaction has been a recurring theme in housing research. The relationship between housing and health has likewise been the focus of considerable work, with early studies exploring the impact upon physical health of relocation from slum neighborhoods to new government housing. Subsequent research turned to issues of psychological adjustment and mental health, in particular the studies of high-rise housing conducted in Great Britain through the 1960s and more recently in Hong Kong and Singapore.

The interior of the dwelling unit has been the focus of limited study by home economics and interior design researchers. Some more recent investigations have begun to look in a detailed fashion at patterns of behavior as they occur within the home environment. Finally it should be noted that different populations — the poor, the elderly, students — have served as the subjects of much housing research.

In response to these various themes in the housing literature, this chapter is organized in four sections. The first and largest includes the bulk of the work on housing effects and the impact of building layout, height, density, quality, and so forth. The second and third sections include materials on student housing and housing for the elderly, reflecting the on-going focus on these two populations and their particular environmental requirements. The final section deals with the area of residential satisfaction. It reflects a shift from a psychological and social-psychological perspective to a more sociological and geographical approach to environment-behavior research as well as a shift to more aggregate environmental variables. As such it represents a bridge to much of the material in Chapter 12 focusing on the larger scale urban environment, as well as the work on behavioral demography in Chapter 14.

a. Housing Effects

1075. Aiello, J. R., & Baum, A. RESIDENTIAL CROWDING AND
         DESIGN. New York: Pergamon Press, 1979. 252 pp.

A review of residential crowding research conducted over the past decade. Effects upon health and education are considered, as is the role of architectural design as a mediating factor.

1076. Becker, F. D. HOUSING MESSAGES. Stroudsburg, Pa.:
         Dowden, Hutchinson & Ross, 1977. 142 pp.

Housing is analyzed as a medium of communication
presenting images and conveying meanings connected to
occupants' feelings, aspirations and experiences. Issues
considered include the importance of personalization,
participation and the housing management process.

1077. Chang, D. Social and psychological aspects of
          housing: A review of the literature. Monticello,
          Ill.: Council of Planning Librarians, 1974.
          (Exchange Bibliographies, no. 557). 20 pp.

Includes 100+ citations along with a review of the
literature as it relates to such variables as density,
privacy and territoriality.

1078. Conway, D. J. (Ed.) HUMAN RESPONSE TO TALL
          BUILDINGS. Stroudsburg, Pa.: Dowden, Hutchinson
          & Ross, 1977. xii, 362 pp.

Based upon a conference of the same title held in 1973.
Papers presented focused on the relation of tall buildings
to their neighborhood, the livability of housing in tall
buildings, and response to emergency in such settings.

1079. Cooper, C. C. EASTER HILL VILLAGE: SOME SOCIAL
          IMPLICATIONS OF DESIGN. New York: The Free
          Press, 1975. xxiv, 337 pp.

A case study of a low-income housing project, including
analyses of space utilization both within and outside the
home. Success in achieving the original design objectives
which guided the project architects is evaluated, and an
extensive series of design recommendations based upon the
study are presented.

1080. Festinger, L., Schachter, S., & Back, K. SOCIAL
          PRESSURES IN INFORMAL GROUPS. Stanford, Calif.:
          Stanford University Press, 1963. 197 pp.

A classic study of "spatial ecology and group
formation," conducted in post-war student housing. Among
this group of highly homogeneous residents, sociometric
choices were inversely proportional to both physical and
functional distance, as well as related to design features
of the apartment buildings.

1081. Gittus, E.   FLATS, FAMILIES AND THE UNDER-FIVES.
          London: Routledge & Kegan Paul, 1976. xvi, 269
          pp.

Part I reviews the findings of a survey with over 340
families with under-five children living in various types of
subsidized housing ranging from high-rises to houses. Part
II considers the rationale for and patterns of development
of high-rise flats during the 1960's and Part III considers
the provision of pre-school and day-care facilities for
children whose home environment may be overly constraining.

1082. Hare, E. H., & Shaw, G. K.   MENTAL HEALTH ON A NEW
          HOUSING ESTATE. London: Oxford Universities
          Press, 1965. 135 pp.

A comparison of the mental health of 1,900 residents of
a post-war British housing development with that of 1,250
residents of an older quarter of the same city. For the
majority of mental health indices no differences were found
between the two groups.

1083. Hassan, R. FAMILIES IN FLATS: A STUDY OF LOW INCOME
          FAMILIES IN PUBLIC HOUSING. Singapore: Singapore
          University Press, 1977. 249 pp.

A study of the daily lives of over 400 low income
families in high-rise public housing in Singapore. Findings
deal with social impacts of relocation, neighborhood
interaction residential satisfaction, and social-
psychological implications of high-rise, high-density
living.

1084. Jephcott, P. HOMES IN HIGH FLATS: SOME OF THE HUMAN
       PROBLEMS INVOLVED IN MULTI-STORY HOUSING.
       Edinburgh: Oliver & Boyd, 1971. 181 pp.

       Interviews with over 600 households residing in high-
rise buildings in Glasgow revealed a number of negative
social impacts, particularly in terms of social isolation
and restrictions in personal choice and self-expression.
Such housing was especially ill-suited to families with
young children.

1085. Moore, W. THE VERTICAL GHETTO: EVERYDAY LIFE IN AN
       URBAN PROJECT. New York: Random House, 1969.
       265 pp.

       A participant observer's impressions of life in urban,
high-rise public housing. Chapter 3 focuses on the
residential environment.

1086. Newman, O. DEFENSIBLE SPACE: CRIME PREVENTION
       THROUGH ENVIRONMENTAL DESIGN. New York:
       Macmillan, 1972. xvii, 264 pp.

       The concept of "defensible space" is developed as an
alternative to traditional security measures as a means of
deterring crime in multi-family urban housing. Qualitative
and quantitative analyses of housing developments in 15
cities are presented, focusing on the relationships between
physical design variables, possibilities for informal
surveillance of public spaces, social contact, and the
occurrence of crime.

1087. Pollowy, A. M. THE URBAN NEST. Stroudsburg, Pa.:
       Dowden, Hutchinson & Ross, 1977. xiv, 182 pp.

An examination of the young child's development as it
relates to the residential milieu, including house or
apartment, the street and planned play spaces. A final
section provides guidelines for both new designs and
improvements in existing environments.

1088. Reynolds, I., & Nicholson, C. THE ESTATE OUTSIDE THE
        DWELLING: REACTIONS OF RESIDENTS TO ASPECTS OF
        HOUSING LAYOUT. London: Department of the
        Environment, 1972. 130 pp.

Reviews the findings of a survey conducted in six
government housing sites with over 1,300 respondents.
Findings deal with housing preferences and desire to move,
maintenance and vandalism, children's play, privacy, multi-
story living and related issues.

1089. Stevenson, A., Martin, E., & O'Neill, J. HIGH
        LIVING: A STUDY OF FAMILY LIFE IN FLATS. New
        York: Cambridge University Press, 1967. 170 pp.

A descriptive study of life in an Australian housing
estate, including residents of both walk-up and high-rise
buildings. Primary shortcomings of the flats were in
meeting the needs of families with children.

1090. Stewart, W. F. R. CHILDREN IN FLATS: A FAMILY
        STUDY. London: National Society for the
        Prevention of Cruelty to Children, 1970. 32 pp.

Comparisons of feelings of loneliness among mothers
living in apartments revealed levels approximately six times
higher in corridor-access as opposed to balcony access
apartment buildings.

1091. Tremblay, K. R. The sociology of housing: An
        updated bibliography. Monticello, Ill.: Vance
        Bibliographies, 1979. (Architecture Series, No.
        A-151) 43 pp.

A topically organized, non-annotated bibliography of over 400 entries. Areas covered include social and psychological effects of housing, housing quality, housing for special populations, and housing policy.

1092. Wilner, D. M. et al. THE HOUSING ENVIRONMENT AND
       FAMILY LIFE: A LONGITUDINAL STUDY OF THE EFFECTS
       OF HOUSING ON MORBIDITY AND MENTAL HEALTH.
       Baltimore: Johns Hopkins University Press, 1962.
       338 pp.

An experimental study involving two groups of 500 families; the experimental group moved from a slum neighborhood into new housing while the control group remained in the same setting. Findings range from solid confirmation of differences in morbidity among children to more modestly suggestive findings regarding the impact of housing on social adjustment.

1093. Zeisel, J., & Griffin, M. CHARLESVIEW HOUSING: A
       DIAGNOSTIC EVALUATION. Cambridge, Mass.:
       Graduate School of Design, Harvard University,
       1975. 131 pp.

Provides a model and methods for conducting post-occupancy evaluation studies and presents findings from one such analysis of a 212 unit housing development. Research questions focused on issues of orientation, territory and privacy.

1094. Altman, I., Nelson, P. A., & Lett, E. E. The ecology
       of home environments. Catalog of selected
       documents in psychology. Washington, D. C.:
       American Psychological Association, 1972, 2 (Ms.
       #421).

Utilizing a detailed survey, stable patterns of space
utilization within the home environment were identified.
Territoriality was found to emerge in the use of bedrooms
and closed doors were recognized and respected as privacy
markers.

1095. Amick, D. J., & Kviz, F. J. Social alienation in
       public housing: The effects of density and
       building type. Ekistics, 1975, 39, 118-120.

Findings from six public housing sites in Chicago
indicated that high social integration, as indicated by a
low level of alienation, was associated with both a low
ratio of dwelling density to site density and low-rise as
opposed to high-rise residence.

1096. Angrist, S. S. Dimensions of well-being in public
       housing families. Environment and Behavior, 1974,
       6, 495-516.

This study explored the relationships between the
physical, social and managerial characteristics of two
public housing projects and five dimensions of well-being of
residents of these settings. Building type (low-rise versus
high-rise) and level of maintenance were found to be
significant predictors of well-being.

1097. Athanasiou, R., & Yoshioka, G. A. The spatial
       character of friendship formation. Environment
       and Behavior, 1973, 5, 43-85.

Friendship in a large, socially heterogeneous community
was found to be a function of both propinquity and
similarity. Similarity in life-cycle stage was important
for friendship formation regardless of distance to friends,
while social-status similarity was important to friendships
among immediate neighbors.

1098.   Bechtel, R. B.   The public housing environment: A
        few surprises.   In #36 (EDRA-3), vol. 1, sec. 13,
        pp. 1-9.

    Application of behavior setting analysis to a large
public housing project in a major northeastern city revealed
a low level of participation and a division among residents
of the community, a lack of space for community use, and a
shifting of the richest settings to the periphery of the
site.   Implications for management and redesign are
presented.

1099.   Beck, R. J., & Teasdale, P.   Dimensions of social
        life style in multiple dwelling housing.   In #1468
        (Esser & Greenbie), pp. 225-251.

    A cross-cultural comparison of the social lifestyles of
75 French-Canadian and English-Canadian families.   Findings
deal with patterns of neighboring, entertaining and mutual
help as well as the impact of life cycle factors and
proximity.   Accompanying design recommendations focus on
private and public spaces adjacent to the dwelling,
community equipment, and mixing of family types.

1100.   Becker, F. D.   The effect of physical and social
        factors on residents' sense of security in multi-
        family housing developments.   Journal of
        Architectural Research, 1975, 4, 18-24.

    Residents of multi-family housing more often attributed
a sense of security to the presence of guards rather than to
design features.   The number of residents' "good friends"
was significantly related to perceived security.

1101.   Bromley, R.   Households suited to high flats.   Town
        and Country Planning, 1979, 48, 46-48.

Contrary to popular stereotypes, over 270 residents of
high-rise blocks were found to be relatively happy with
their housing accommodations. The greatest level of
dissatisfaction occurred in households with children.

1102.   Bryant, D., & Knowles, D. Social contacts on the
        Hyde Park Estate, Sheffield. Town Planning
        Review, 1974, 45, 207-214.

Within a large housing complex of innovative design,
incorporating upper-level access "streets," residents were
found to have lower levels of contact among themselves than
with individuals in the surrounding community.

1103.   Canter, D., & Thorne, R. Attitudes to housing: A
        cross-cultural comparison. Environment and
        Behavior, 1972, 4, 3-32.

Contrary to expectations at the outset of the study,
beginning architectural students in Australia preferred old
terrace/row housing, while Scottish students preferred the
traditional individual bungalow style typical of Australia.
It is suggested that the methodology employed, utilizing
slides and semantic differential scales, can be utilized in
different cultural contexts.

1104.   Chave, S. P. Mental health in Harlow New Town.
        Journal of Psychosomatic Research, 1966, 10,
        38-44.

Residents of a British New Town were found to exhibit
less major psychiatric illness than the general population,
but comparable levels of neurosis. There were no
indications of a neurosis characteristic of the new
community, either in manifestation or prevalence.

1105.   Churchman, A., & Herbert, G. Privacy aspects in the
        dwelling. Journal of Architectural Research,
        1978, 6(3), 19-27.

A theoretical analysis of design implications of privacy within the home environment. Privacy is conceptualized as the management of interpersonal interaction and the flow of information.

1106. Conway, J., & Adams, B. The social effects of living
         off the ground. Habitat International, 1977, 2,
         595-614.

A review of recent research into the social effects of housing, and particularly those studies conducted by the Social Research Division of the British Department of the Environment. Benefits and problems of high-rise living for different social groups are considered.

1107. Cooper, C. C. Fenced back yard - unfenced front yard
         - enclosed back porch. Journal of Housing, 1967,
         24, 268-274.

Interviews and observation at Easter Hill Village support the functional and psychological importance of private back yards. Front porches were perceived as contributing some uniqueness to the project and also served a variety of child care, storage, and social functions.

1108. Egolf, B., & Herrenkohl, R. C. The influence of
         familiarity and age factors on responses to
         residential structures. In #1078 (Conway), pp.
         208-217.

Based upon responses to slides of different housing types, preferences were found to be related to subjects' familiarity with a given type of housing but not to subjects' age.

1109. Eoyang, C. K. Effects of group size and privacy in
       residential crowding. Journal of Personality and
       Social Psychology, 1974, 30, 389-392.

   Among a sample of residents of identical house trailers,
number of occupants per unit accounted for greater variance
in living space satisfaction than did degree of privacy
within units.

1110. Essen, J., Fogelman, K., & Head, J. Childhood
       housing experiences and school attainment. Child
       Care, Health and Development, 1978, 4, 41-58.

   Data from a longitudinal national sample indicate that
children who have lived in crowded conditions, without basic
sanitary facilities, or in government housing had lower
scores on tests of school attainment than did other children
from similar social backgrounds.

1111. Fanning, D. M. Families in flats. British Medical
       Journal, 1967, 18, 382-386.

   Children and mothers in apartments had less access to
outdoor space for play and social contact. Respiratory
disease rates were found to be twice those of children
living in homes. Mothers in apartments were found to have
higher rates of psycho-neurotic disorders.

1112. Gillis, A. R. High-rise living and psychological
       strain. Journal of Health and Social Behavior,
       1977, 18, 418-431.

   Gender and floor level of residents' apartments were
found to interact in predicting psychological strain. For
women, a direct and significant relationship between floor
level and strain emerged, while an inverse but non-
significant relationship emerged for males.

1113.   Greenberg, J., & Greenberg, C. I.   A survey of
            residential responses to high-rise living.   In
            #1078 (Conway), pp. 168-174.

    Responses to a set of semantic differential descriptors
indicated upper-floor residents of an apartment building to
perceive their apartments as significantly quieter, more
efficient, more functional but less secure than did
residents of lower floors.

1114.   Haber, G. M.   The impact of tall buildings on users
            and neighbors.   In #1078 (Conway), pp. 45-57.

    The view of scenery was the feature of tall buildings
most liked by a sample of university students.   Waiting for
elevators, lack of greenery, and fear of fire were the least
liked features.   Ten percent of respondents indicated that
they would not live in a tall building and 2% that they
would not work in one.

1115.   Kasl, S. V.   Effects of housing on mental and
            physical health.   Man-Environment Systems, 1974,
            4, 207-226.

    A review of the literaure, under six headings:
satisfaction with housing and neighborhood, urban ecological
analysis, parameters of housing and neighborhood, voluntary
rehousing, involuntary relocation, and proximate
environment.

1116.   Kreisberg, L.   Neighborhood setting and the isolation
            of public housing tenants.   Journal of the
            American Institute of Planners, 1968, 34, 43-49.

    Residence in public housing represents a barrier to
interaction with the surrounding community unless tenants
are largely drawn from the local population, do not develop
their own strong community, and are not separated by marked
physical barriers.

1117. Ladd, F. Black youths view their environment: Some
       views on housing. Journal of the American
       Institute of Planners, 1972, 38, 108-116.

       A description and analysis of the residential
environmental features which were meaningful to 60
adolescent boys. Their responses revealed both a socio-
economic and a psychological discrepancy between current
home environment and that to which they aspired.

1118. Marcus, C. C. Children's play behavior in a low-rise
       inner-city housing development. In #38 (EDRA-5),
       v. 12, pp. 197-211.

       Low-rise housing was perceived by parents to be a
substantially more attractive and safer environment for
children than a high-rise complex. More than twice as many
mothers allowed their children to play outside alone in the
low-rise as compared to the high-rise housing.

1119. Martin, A. E. Environment, housing and health.
       Urban Studies, 1967, 4, 1-21.

       A review article exploring relationships between
environment, housing and health from an epidemiological
perspective.

1120. McCarthy, D., & Saegert, S. Residential density,
       social overload and social withdrawal. Human
       Ecology, 1978, 6, 253-272.

       Residents of high-rise public housing towers reported
seeing people more often and feeling more crowded in their
building, felt less control, privacy and security, and were
less socially active and satisfied than residents of low-
rise buildings within the same project.

1121. Mitchell, R. E. Ethnographic and historical
          perspectives on relationships between physical and
          socio-spatial environments. Sociological
          Symposium, 1975, 14, 25-40.

     It is argued that public policy as it affects housing
design and density is subject to substantial cultural
biases. Ethnographic evidence is cited to illustrate the
positive rather than negative potential of multi-family,
high-density living.

1122. Mucheno, M. D., Levine, J. P., & Palumbo, D. J.
          Television surveillance and crime prevention:
          Evaluating an attempt to create defensible space
          in public housing. Social Science Quarterly,
          1978, 58, 647-656.

     The introduction of a closed circuit television
surveillance system, as proposed by Newman (item #1086), had
no significant impact upon occurrence of crime or fear of
crime within a public housing project. Questions for a
theory of defensible space are raised.

1123. Mullins, P., & Robb, J. H. Residents' assessment of
          a New Zealand public-housing scheme. Environment
          and Behavior, 1977, 9, 573-624.

     Provision of single-family housing, contact with
neighbors, and localization of friends and kin contributed
to a largely positive evaluation of both dwelling and
residential environments.

1124. Neutra, R., & McFarland, R. A. Accident epidemiology
          and the design of the residential environment.
          Human Factors, 1972, 14, 405-420.

Reviews a variety of studies which analyzed the
relationships between housing and incidence of accidents.
It is suggested that better design can reduce accident
rates; a variety of preventative measures are discussed.

1125.   Pollowy, A. M.   Children in high-rise dwellings.   In
        #1078 (Conway), pp. 149-160.

An excerpt from THE URBAN NEST (item #1087), presenting
a number of design guidelines based upon both the child
development literature and studies of children's play.

1126.   Rainwater, L.   Fear and the house-as-haven in the
        lower class.   Journal of the American Institute of
        Planners, 1966, 32, 23-31.

Observations and interviews over a 10-year period
suggest that the lower class seeks shelter from both human
and non-human threats in their housing.   Residents of public
housing place great emphasis on both privacy and security.

1127.   Reynolds, K., & Nicholson, D.   Living off the ground.
        Architects Journal, 1967, 150, 459-470.

Studies of over 1,000 households in six housing estates
throughout Britain revealed no differences in satisfaction
between those living in high-rise and low-rise housing.
There were differences in the play habits of children, and
buildings over 20 stories in height were regarded less
favorably.

1128.   Richman, N.   The effects of housing on pre-school
        children and their mothers.   Developmental
        Medicine and Child Neurology, 1974, 16, 53-58.

Among a sample of 75 families with at least two children
under five years, housing dissatisfaction was substantially

higher among tenants in flats in both high-rise and low-rise buildings than among those living in houses. In addition, there were more complaints regarding loneliness and depression among mothers living in flats.

1129. Rothblatt, D. W. Housing and human needs. Town Planning Review, 1971, 42, 130-144.

The impact of large scale high-density housing on family, belongingness, esteem, and independence "needs" was assessed in two publicly assisted projects. Conclusions favor the provision of low-rise structures for families with children.

1130. Saegert, S., & Winkel, G. The home: A critical problem for changing sex roles. In G. R. Wekerle, R. Peterson, & D. Morley (Eds.), NEW SPACE FOR WOMEN. Boulder, Co.: Westview Press, 1980. Pp. 41-63.

Women were found to invest more labor in the home and to place greater value on the home expressing their personalities than did men. Women in the city and men in the suburbs tend to value home and work equally, whereas urban men emphasize work and suburban women focus on the home.

1131. Scheflen, A. E. Living space in an urban ghetto. Family Process, 1971, 10, 429-450.

An analysis of the organization of home space in urban black and Puerto Rican families. While commonalities were found in territorial behavior, differences emerged in extent of home utilized and organization of space by activity or age group.

1132. Schorr, A. L. Housing and its effects. In R. Gutman & D. Popenoe (Eds.), NEIGHBORHOOD, CITY AND

METROPOLIS. New York: Random House, 1970. Pp.
/09-729.

Based upon a review of the literature, Schorr concludes
that the impact of physical housing is greater than is
commonly assumed. Home and neighborhood are considered both
separately and together in terms of their impact upon self-
perception, health, satisfaction, stress, social and family
relationships and various effects of crowding.

1133. Smith, R. H., et al. Privacy and interaction within
       the family as related to dwelling space. Journal
       of Marriage and the Family, 1969, 31, 559-566.

An observational study of living patterns of 20 families
from four different stages of the life cycle, each of whom
lived in an actual house for two weeks. Division and
allocation of space, privacy-interaction orientation of
family members, husband's occupation, and stage in life
cycle all influenced the relationship between space and
interaction in the study dwelling.

1134. Steidel, R. E. Difficulty factors in homemaking
       tasks: Implications for environmental design.
       Human Factors, 1972, 14, 471-482.

Among a sample of 208 wives, the impact of the immediate
physical environment upon their ability to effectively
perform food, shelter, clothing, and family/personal tasks
was mentioned in roughly 50% of responses; the environment
was not mentioned in regard to management tasks however.

1135. Tulkin, S. R., & Kagan, J. Mother-child interaction
       in the first year of life. Child Development,
       1972, 43, 31-41.

Working class homes were found to be characterized by
more noise, greater crowding, more interaction with many
adults, and more time spent in viewing television.

1136. Wachs, T. The measurement of early intellectual
         functioning. In C. Meyers, R. Eyman & G. Tarjan
         (Eds.), SOCIO-BEHAVIORAL STUDIES IN MENTAL
         RETARDATION. Washington, D. C. American
         Association on Mental Deficiency, 1973.

    The availability of "stimulus shelter," a room to which
a child could escape intense social stimulation, was
positively and strongly related to a variety of cognitive
measures and served as a predictor of later cognitive
development.

1137. Walter, M. The territorial and the social:
         Perspectives on the lack of community in high-
         rise/high density living in Singapore. Ekistics,
         1978, 45, 236-242.

    An analysis of patterns of social interaction and social
support in high-rise housing and in the traditional Malay
village.

1138. Wekerle, G., & Hall, E. High rise living: Can the
         same design serve young and old? Ekistics, 1972,
         33, 186-191.

    Young single residents of an urban housing development
placed a priority on social desires and were relatively
indifferent to features of the physical environment which
did not contribute to these social goals. A smaller group
of older tenants placed more emphasis on both convenience
and the quality of the physical setting.

1139. Williamson, R. C. Socialization in the high-rise: A
         cross-national comparison. Ekistics, 1978, 45,
         122-130.

A study of residents of both high-rise and more
traditional housing structures in Germany and Italy.
Findings with regard to security, social networks, and
socialization of children revealed fewer differences than
had been anticipated.

1140. Wilner, D., & Walkeley, R. P. Effects of housing on
          health and performance. In L. Duhl (Ed.), THE
          URBAN CONDITION. New York: Basic Books, 1963.
          Pp. 215-228.

A review of 40 studies of the relationship between
housing type and physical and social pathologies indicated
that material and familial considerations, as well as poor
housing, contributed to maladjustment.

1141. Yancey, W. L. Architecture, interaction, and social
          control: The case of a large-scale public housing
          project. Environment and Behavior, 1971, 3, 3-21.

Ethnographic research in Pruitt-Igoe (a public housing
project in St. Louis) suggests that the lack of semi-public
spaces between apartments had an atomizing effect on the
informal social networks essential to working and lower
classes. This resulted in a loss of social support,
protection, and informal social control.

1142. Yeung, Y. High-rise, high-density housing: Myths
          and reality. Habitat, 1977, 2, 587-594.

Critically examines four assumptions about high-rise,
high-density residential environments: that high-rise living
interferes with a sense of community, that it has adverse
psychological and social effects, that high-rise housing
means high-density living conditions, and that criteria used
to measure the effects of such conditions are applicable on
a cross-national basis. The research literature is found to
be wanting in regard to unequivocal support for any of these
"myths."

1143. Zito, J. M. Anonymity and neighboring in an urban,
       high-rise complex. Urban Life and Culture, 1974,
       3, 243-263.

    Despite judging other tenants as being similar to
themselves, residents of an urban high-rise avoided any
contact with neighbors beyond perfunctory greetings.

1144. Cagle, L. T. Neighboring in public housing projects:
       A study of the relationships between status
       homophily and propinquity. Dissertation Abstracts
       International, 1969, 30, 3384.

1145. Champommier, J. G. Emotional stress perceptions,
       spatial behavior, and the socio-physical
       environment: A study of a public housing project
       in Northeast Los Angeles. Dissertation Abstracts
       International, 1976, 37-A, 1838.

1146. Chenoweth, R. E. The effects of territorial marking
       on residents of two multifamily housing
       developments: A partial test of Newman's theory
       of defensible space. Dissertation Abstracts
       International, 1978, 38-B, 5088-5089.

1147. Humphries, G. M. Values, satisfaction, aspirations
       and goal commitment among multiunit housing
       residents. Dissertation Abstracts International,
       1976, 37-A, 3213.

1148. Rohe, W. M. Attitudinal, behavioral and health
       correlates of residential density and housing
       type. Dissertation Abstracts International, 1979,
       39-B, 4109.

See also: citations 96, 124, 125, 331, 337, 446, 580, 1441.

b. Student Housing

1149. Baum, A., & Valins, S. ARCHITECTURE AND SOCIAL
        BEHAVIOR: PSYCHOLOGICAL STUDIES IN SOCIAL
        DENSITY. Hillsdale, N.J.: Lawrence Erlbaum
        Associates, 1977. 112 pp.

A summary of a series of field and laboratory studies of
the experiential and behavioral consequences of two
different residence hall environments. One dormitory was of
the standard corridor arrangement with bath and lounge
shared by 34 people, while the other grouped six students in
suites. Despite the comparable densities of the two
settings, residents of the corridor dorm experienced a
syndrome of social and behavioral withdrawal, seemingly
reflecting their inability to regulate social experiences.

1150. Heilweil, M. (Ed.) Student housing, architecture,
        and social behavior. Environment and Behavior
        1973, 5, 375-513.

A special issue with contributions concerning suite
versus corridor arrangements in dormitories, the impact of
overall density, and the role of student-built and minimal-
cost housing.

1151. Van der Ryn, S., & Silverstein, M. DORMS AT
        BERKELEY: AN ENVIRONMENTAL ANALYSIS. New York:
        Educational Facilities Laboratories, 1967. 89 pp.

A pioneering study in post-occupancy evaluation focused
on high-rise dormitories. Student dissatisfaction was
related to lack of personal choice and control, provision of

innappropriate spaces, and a failure to accommodate varying
student life styles.

1152. Alfert, E. Housing selection, need satisfaction, and
      dropouts from college. Psychological Reports,
      1966 19, 183-186.

   Students in private rooms and boarding houses felt more
isolated and had a higher dropout rate than did comparable
students living in dormitories, fraternaties and sororities,
apartments, and cooperatives.

1153. Ankele, C., & Sommer, R. The cheapest apartments in
      town. In #1150 (Heilweil), pp. 505-513.

   Interviews in an ill-kept but inexpensive apartment
complex suggest a role for such housing. For its relatively
transient population, low rent and privacy were more
important than high levels of social interaction and
amenities.

1154. Baird, L. L. The effects of college residence groups
      on students' self-concepts, goals, and
      achievements. Personnel and Guidance Journal,
      1969, 47, 1015-1021.

   In a national sample of sophomores, students living on-
campus were found to be more active in university activities
than those living off-campus. Little difference was
observed in academic variables.

1155. Baron, R., et. al. Effects of social density in
      university residential halls. Journal of
      Personality and Social Psychology, 1976, 34,
      434-446.

Residents of triple occupancy rooms which had been
initially designed to accomodate two people expressed
greater feelings of crowding, perceived less control over
room activities, expressed more negative personal attitudes,
and experienced a more negative room ambience than did those
students sharing similar rooms with only one other person.

1156. Corbett, J. A. Are suites the answer? In #1150
       (Heilweil), pp. 413-419.

Questionnaire responses from residents of four person
suites in a university dormitory suggested strong individual
differences in evaluations of this living arrangement.
Satisfaction was increased by self-selection of and cooperation among roommates.

1157. Corbett, J. A. Student-built housing as an
       alternative to dormitories. In #1150 (Heilweil),
       pp. 491-504.

Includes a brief history and evaluation of student built
dome housing. The domes were rated as equal to or better
than dormitories and apartments in most respects, and met
their builders aspirations of community and privacy.

1158. Corbett, J. Student-built housing: A second visit.
       Journal of Architectural Research, 1977, 6, 36-39.

A follow-up evaluation of student-built and maintained
housing three years after its completion. Satisfaction
remained very high and group spirit, while not as strong as
during the construction period, was still very positive.

1159. Davis, G., & Roizen, R. Architectural determinants
       of student satisfaction in college residence
       halls. In #35 (EDRA-3), pp. 28-44.

Based upon questionnaire responses from students in 43
residence halls on eight campuses, it is concluded that
individual environmental features have little to do with
overall satisfaction. Overall impression of the dormitory
is the best predictor, with those students living in the
least conventional dormitories being the most satisfied.

1160. Duvall, W. H. Student-staff evaluations of residence
      hall environments. The Journal of College Student
      Personnel, 1969, 10, 52-58.

A Residence Hall Environment Index was developed and
employed to assess student resident satisfaction in terms of
group living, student government, and physical facilities.

1161. Gerst, M. S., & Moos, R. H. The social ecology of
      university student residences. Journal of
      Educational Psychology, 1972, 63, 513-525.

Reviews the development, standardization, and
utilization of the University Residence Environment Scales.
Some potential relationships between architectural variables
and social climate are proposed.

1162. Gerst, M. S., & Sweetwood, H. Correlates of
      dormitory social climate. In #1150 (Heilweil),
      pp. 440-464.

The University Residence Environment Scale (URES)
revealed only minor differences between two dormitory types
differing in architectural design. Students who described
the social climate of their dormitories or suites in more
positive terms had a more extended and intimate network of
friendships, a more positive set of subjective moods, and
perceived their building as more attractive and interesting.

1163. Heilweil, M. The influence of dormitory architecture
      on resident behavior. In #1150 (Heilweil), pp.
      377-412.

A review of the impact of architectural design on
privacy, social relations, study behavior and
individualization. Alternative philosophies of education
and student life and their relationship to environmental
design are also considered.

1164. Holahan, C. J., & Wilcox, B. Residential
       satisfaction and friendship formation in high and
       low rise student housing: An interactional
       analysis. Journal of Educational Psychology,
       1978, 70, 237-241.

   Residents of low-rise dormitories were found to be
significantly more satisfied and to establish more
dormitory-based friendships than residents of high-rise
buildings. Satisfaction and friendship formation were also
related to both sex and social competence of residents.

1165. Menne, J. M., & Sinnett, E. R. Proximity and social
       interaction in residence halls. Journal of
       College Student Personnel, 1971, 9, 26-32.

   Based upon sociometric analyses, adjacency within a
dormitory predicted reciprocal choices of friends and
helpers. Roommates were most likely to be chosen, followed
by near neighbors, others in the corridor, and finally
others. Adjacency did not predict non-reciprocal choices.

1166. Moos, R. H. Social environments of university
       student living groups: Architectural and
       organizational correlates. Environment and
       Behavior, 1978, 10, 109-126.

   Based upon analyses of 87 student living units,
proportion of single rooms and location on campus were the
physical variables most strongly related to social climate.

1167. Preiser, W. F. E. Behavioral design criteria in
       student housing. In #34 (EDRA-1), pp. 243-259.

    A two-stage study involving the construction and scaling
of attitude statements related to dormitories. Statements
were categorized in terms of physical-environmental,
psychologival, sociological, and regulatory features. The
study concludes with a number of recommended design
criteria.

1168. Sommer, R. Student reactions to four types of
       residence halls. Journal of College Personnel,
       1968, 9, 232-237.

    Surveys were distributed to residents of small cluster
dormitories, high-rise buildings, apartments, and converted
barracks. Findings regarding satisfaction with social life,
isolation from campus, identification with floor group, and
overall satisfaction were reviewed.

1169. Sommer, R. Study conditions in student residences.
       Journal of College Student Personnel, 1969, 10,
       270-274.

    Data gathered from visits to over 1,000 individual
student rooms on eleven university and college campuses
indicated that a student was more likely to be studying if
his/her roommate was present and studying than if the
roommate was away. While most studying took place at desks,
a sizable amount took place on beds.

1170. Stokols, D., Ohlig, W., & Resnick, S. Perception of
       residential crowding, classroom experiences, and
       student health. In #1468 (Esser & Greenbie), pp.
       89-111.

    Data gathered from 27 university students indicated that
unfavorable reactions to physical and social aspects of non-

residential settings were associated with perceived crowding
in and negative feeling abouts their dormitory environment.
Number of visits to physicians and academic performance over
one term were related to ratings of residential crowding,
social climate and physical conditions.

1171. Wilcox, B. L., & Holahan, C. Social ecology of the
       megadorm in university student housing. Journal
       of Educational Psychology, 1976, 68, 453-458.

In a comparison with residents of smaller dormitories,
students in high-rise "megadorms" reported less involvement,
less emotional support, less control of events, and less
order and neatness.

1172. Johnson, W. E. Student perceptions of residence hall
       environments. Dissertation Abstracts
       International, 1969, 30-A, 1050-1051.

See also: citations 413, 469, 557, 874, 1381, 1383.

c. Housing for the Elderly

1173. Carp, F. M. A FUTURE FOR THE AGED: VICTORIA PLAZA
       AND ITS RESIDENTS. Austin, Tx., University of
       Texas Press, 1976. xv, 287 pp.

One year after a voluntary move to public housing for
the elderly, residents were higher in morale, perceived

health, social activity and housing satisfaction than were a
group of unsuccessful applicants for the same building who
continued to live in the community.

1174.   Koch, J. E. Housing and the aged, 1964-1975: A
          selected bibliography. Monticello, Ill.: Council
          of Planning Librarians, 1976. (Exchange
          Bibliographies, No. 1028) 60 pp.

     An extensive but non-annotated bibliography with
emphasis on residential settings for the elderly other than
nursing homes.

1175.   Lawton, M. P. ENVIRONMENT AND AGING. Belmont,
          Calif.: Brooks-Cole, 1980. xii, 186 pp.

     A concise summary of current theory and research within
the area of environments and aging. An integrative
theoretical model is presented along with chapters on
residential patterns among older persons, congregate and
institutional housing, and guidelines for increasing the
impact of environmental research.

1176.   Brody, E. M. Community housing for the elderly. The
          Gerontologist, 1978, 18, 121-128.

     Among a group of applicants to community housing for the
elderly, mortality rates over three years were significantly
higher for those who did not move into this setting. The
"immobilization effect" is advanced as a potential
explanation for these findings.

1177.   Brody, E. M., Kleban, M. H., & Liebowitz, B.
          Intermediate housing for the elderly:
          Satisfaction of those who moved in and those who
          did not. The Gerontologist, 1975, 15, 350-356.

Older persons who relocated into semi-independent "community housing" were more satisfied with their housing, level of friendship, and enjoyment of life than were older persons who relocated into other kinds of housing or who did not move.

1178. Carp, F. M. Impact of improved living environment on health and life expectancy. The Gerontologist, 1977, 17, 242-249.

Data gathered 8 years after Carp's initial study (item #1173) demonstrate the continuing benefits of relocation, typically from substandard housing, to a new high-rise for the elderly. Significant differences were found for both self and interviewer perceptions of health, frequency of medical contact, and mortality rates.

1179. Cluff, P. J., & Campbell, W. H. The social corridor. The Gerontologist, 1975, 15, 516-523.

An in-use evaluation of a 150-bed home for the aged indicated a relationship between the configuration of building corridors and social interactions, activity patterns, and resident satisfaction. Satisfaction declined with increasing distance between resident rooms and commonly used facilities.

1180. Cranz, G., & Schumacher, T. L. The impact of high-rise housing on older residents. In #1078 (Conway), pp. 194-207.

Data gathered in eight government supported independent living projects indicated a negative impact of high-rise living only with respect to reported degree of boredom, amount of activity and numbers of other residents known. For various other variables - preferred location, use of communal space, security and privacy - results were either equivocal or the consequence of other factors such as neighborhood context.

1181. Jones, D. C. Spatial proximity, interpersonal
conflict, and friendship formation in the
intermediate care facility. The Gerontologist,
1975, 15, 150-154.

Study of over 400 patients in 10 intermediate care
nursing homes revealed that incidents of conflict were more
likely to occur among individuals living in the same or
adjacent rooms. Sustained positive interactions were more
likely to occur among patients who resided at least two
rooms apart from one another.

1182. Larson, C. J. Alienation and public housing for the
elderly. International Journal of Aging and Human
Development, 1974, 5, 224-225.

Levels of alienation among elderly residents of public
housing were lower for those accustomed to living in
multiple family dwellings. Contact with friends and
relatives, and visiting with other tenants were found to
mitigate these effects.

1183. Lawton, M. P. Public behavior of older people in
congregate housing. In #35 (EDRA-2), pp. 372-380.

Within a sample of 15 congregate housing sites, a larger
number of building-centered social activities were related
to a greater proximity to resources in the larger community.
Increased social activity was also related to health, number
of within-building social spaces, and tenants' propensity to
leave their apartment doors ajar.

1184. Lawton, M. P. The relative impact of congregate and
traditional housing on elderly tenants. The
Gerontologist, 1976, 16, 237-242.

Contrasted with elderly tenants relocated to traditional
housing for the elderly, those who moved to congregate
housing with some on-site services showed relative
improvement in morale, available social network, and housing
satisfaction, along with decrements in two measures of
involvement with the external world.

1185.   Lawton, M. P., Brody, E. M., & Turner-Massey, P. The
        relationships of environmental factors to changes
        in well-being. The Gerontologist, 1978, 18,
        133-137.

Based upon a sample of 82 older persons, favorable
neighborhood and residential factors were associated with
positive effects on well-being.

1186.   Lawton, M. P., & Cohen, J. The generality of housing
        impact on the well-being of older people. Journal
        of Gerontology, 1974, 29, 194-204.

After one year, rehoused elderly compared favorably with
community residents in terms of five indices of well-being
including morale and housing satisfaction. The rehoused
elderly were poorer in functional health and no differences
were found in three other indices.

1187.   Lawton, M. P., Nahemow, L., & Teaff, J. Housing
        characteristics and the well-being of elderly
        tenants in federally assisted housing. Journal of
        Gerontology, 1975, 30, 601-607.

Based upon a national probability sample of housing for
the elderly, housing satisfaction was found to be greater in
projects which were smaller in terms of total number of
units; greater building height was associated with lower
housing satisfaction and less neighborhood motility.

1188. Lawton, M. P., & Simon, B. B. The ecology of social
       relationships in housing for the elderly. The
       Gerontologist, 1968, 8, 108-115.

   Residents in housing for the elderly were more likely to
interact socially and maintain friendships with people
living on their own floors and with those living next door
than with those living in locations further away.

1189. Lipman, A. Public housing and attitudinal adjustment
       in old age: A comparative study. Journal of
       Geriatric Psychiatry, 1968, 2, 88-101.

   A sample of over 600 elderly residents of various public
housing projects were found to have higher morale and higher
scores on a variety of indices of successful aging than did
a matched control group of over 100 older persons who had
applied for but not been accepted into public housing.

1190. Moran, R., & Schafer, S. Criminal victimization of
       the elderly in three types of urban housing
       environments. Housing and Society: The Journal
       of the American Association of Housing Educators,
       1978, 5.

   Based upon a survey of elderly persons living in both
age-segregated and other types of housing, less
victimization was found to occur in elderly-only units.

1191. Nahemow, L., & Lawton, M. P. Similarity and
       propinquity in friendship formation. Journal of
       Personality and Social Psychology, 1975, 32,
       205-213.

   A sociometric study conducted in a city housing project
demonstrated residents' daily activity space to be a region
in which they knew virtually everyone; less proximal
friends, however, were likely to be of the same age and

race. There was little evidence that older residents had
fewer, more proximal, or more similar friends.

1192. Nahemow, L., Lawton, M. P., & Howell, S. C. Elderly
        people in tall buildings: A nationwide study. In
        #1078 (Conway), 175-181.

    Results indicate that when other demographic variables
are controlled, the effect of high-rise living on older
people is not very great. Dissatisfaction is highest among
those who have moved from single family homes; primary
concerns are fire safety, elevators and lack of secondary
entries.

1193. Patterson, A. H. Housing density and various quality
        of life measures among elderly urban dwellers.
        Journal of Population, 1978, 1, 203-215.

    The significance of findings that the elderly appear to
be relatively unaffected by conditions of high-density
living is discussed in the light of activity theory,
suggesting a need for higher levels of social interaction in
this age group. A case study of a group of elderly living
in a small Northeast city is described in relation to this
interpretation.

1194. Weber, R. J., Brown, L. T., & Weldon, J. K.
        Cognitive maps of environmental knowledge and
        preference in nursing home patients. Experimental
        Aging Research, 1978, 4, 157-174.

    Differential performance in identifying the locations of
slide-depicted scenes within a nursing home was found to be
influenced by both age of patients and relative visual
distinctiveness of areas of the building. Patients' own
rooms were markedly preferred to common areas.

<u>See</u> <u>also</u>: citations 65, 485, 489, 518, 605.

d. Residential Satisfaction

1195. Francescato, G. et al. RESIDENTS' SATISFACTION IN
      HUD-ASSISTED HOUSING: DESIGN AND MANAGEMENT
      FACTORS. Washington, D. C.: Office of Policy
      Development and Research, U. S. Department of
      Housing and Urban Development, 1979. xiv, 163 pp.

   A review of a five year research program on residential
satisfaction based upon a nationwide sample of 37 housing
developments. Those factors found to contribute most to
satisfaction were satisfaction with other residents,
pleasant apearance, and economic value of housing.

1196. Lansing, J. B., Marans, R. W., & Zehner, R. B.
      PLANNED RESIDENTIAL ENVIRONMENTS. Ann Arbor,:
      Survey Research Center, University of Michigan
      1970. 269 pp.

   A comparison of five highly planned and five less
planned communities matched in terms of dwelling types and
values, ownership, and length of residency. While not
directly related to residential satisfaction, dwelling unit
density influenced such important factors as privacy in the
yard, neighborhood noise level, and adequacy of outdoor
space.

1197. Lev-Ari, A. Residential preferences in metropolitan
      areas. Monitcello, Ill,: Vance Bibliographies,
      1979. (Public Administration Series, No. 385) 8
      pp.

A non-annotated bibliography of approximately 100 items.

1198. Michelson, W. ENVIRONMENTAL CHOICE, HUMAN BEHAVIOR,
        AND RESIDENTIAL SATISFACTION. New York: Oxford
        University Press, 1977. xix, 403 pp.

A five year longitudinal study of patterns of decision
making in housing selection and relocation among residents
of center city and suburban Toronto. The fit between people
and their housing is analyzed in terms of reasons for
moving, expectations, subsequent experiences, and
evaluations.

1199. Zehner, R. B. INDICATORS OF THE QUALITY OF LIFE IN
        NEW COMMUNITIES. Cambridge, Mass.: Ballinger,
        1977. xxxii, 243 p.

Fifteen comprehensively planned new communities and 21
comparison communities were assessed using four scales of
indicators. In terms of dwelling satisfaction and overall
quality of life, responses of residents in the two groups of
communities were quite similar. More significant
differences emerged at neighborhood and community levels

1200. Ahlbrandt, R. S., & Brophy, P. C. Management: An
        important element of the housing environment.
        Environment and Behavior, 1976, 8, 505-526.

Tenant satisfaction is shown to consist of discrete
dimensions which include neighbors, neighborhood cleanliness
and services, the physical unit, security and management; of
these dimensions, management is the most important
determinant of satisfaction.

1201. Bardo, J. W. Dimensions of community satisfaction in
a British New Town. Multivariate Experimental
Clinical Research, 1976, 2, 129-134.

Factor analysis of community-satisfaction questionnaire
data obtained from new-town residents yielded nine factors,
seven relating to the social and two to the physical
environment.

1202. Barrett, F. The search process in residential
location. Environment and Behavior, 1976, 8,
169-197.

Purchase decisions of 380 new home buyers in Toronto
were made after a very short search which covered just a few
homes in a small area. A composite time/area index
indicated more intensive and spatially concentrated
searching in inner-city ethnic neighborhoods.

1203. Bortz, J. Erkundingsexperiment zur Beziehung
zwischen Fassadengestaltung und ihrer Wirkung auf
den Betrachter. Zeitschrift fur experimentelle
und angewandte Psychologie, 1972, 19, 228-281.

Attributes and dimensions of apartment houses, assessed
via a combination of judges' ratings and informational
measures, are correlated with semantic-differential ratings
of the buildings facades, via the application of canonical
correlation methods, revealing consistent predictive value
of particular architectural features, notably the presence
or absence of a balcony.

1204. Brink, S., & Johnston, K. A. Housing satisfaction:
The concept and evidence from home purchase
behavior. Home Economics Research Journal, 1979,
7, 338-345.

Data gathered from new home buyers suggest that housing
satisfaction may be viewed as a function of realization of
housing aspirations, fulfillment of expectations, and degree
of housing improvement. No significant decline in housing
satisfaction over time was observed.

1205.   Campbell, A., Converse, P. E., & Marans, R. W.   The
        residential environment. In A. Campbell & P. E.
        Converse (Eds.), THE QUALITY OF AMERICAN LIFE:
        PERCEPTIONS, EVALUATIONS AND SATISFACTIONS.   New
        York: Russell Sage Foundation, 1976.   Pp.
        217-266.

A revised version of a previous paper by Marans and
Rodgers (item #1212). Satisfaction within this domain is
considered at community, neighborhood, and residential
scales.

1206.   Eng, T. S.   Residential choice in public and private
        housing in Singapore. Ekistics, 1978, 45,
        246-249.

Neighborhood quality (low density, clean and quiet
surroundings, suitability for children), accessibility, and
opportunities for socialization were found to be basic to
choice of a home. These factors were found to cut across
income and current housing situation.

1207.   Francescato, G. et al.   Predictors of residents'
        satisfaction in high-rise and low-rise housing.
        Journal of Architectural Research, 1975, 4(3),
        4-9.

Survey results indicated few significant differences in
satisfaction for residents of high versus low-rise buildings
in urban settings. Privacy, safety and security, and the
immediate neighborhood were important predictors of
satisfaction only for high-rise residents.

1208. Hinshaw, M., & Allott, K. Environmental preferences
        of future housing consumers. Journal of the
        American Institute of Planners, 1972, 38, 102-107.

    While housing preferences of urban youth showed some
specific differences among socio-economic groupings, there
was no real rejection of the pattern of single-family home
ownership.

1209. Jaanus, H., & Nieuwenhuijse, B. Determinants of
        housing preference in a small town. In #1468
        (Esser & Greenbie), pp. 252-274.

    Eight neighborhoods within the study community were
rated in terms of 36 scales reflecting characteristics of
these areas. Three dimensions emerged out of a factor
analysis of these ratings with one factor, dealing with
comfort and privacy, strongly associated with housing
preference. A second factor related to open space showed
little relationship to preference and the third, dealing
with lively urban spaces, showed an inverse relationship.

1210. Lamanna, R. A. Value consensus among urban
        residents. Journal of the American Institute of
        Planners, 1964, 30, 317-323.

    A study of the role of thirteen specific physical and
social attributes of a city as contributors to "livability,"
based on responses of residents of Greensboro, N. C.
Differences related to race, sex, age and socio-economic
status are brought out.

1211. Lansing, J. B., & Marans, R. W. Evaluation of
        neighborhood quality. Journal of the American
        Institute of Planners, 1969, 35, 195-199.

    A survey of over 1 000 people in almost 100
neighborhoods in the Detroit area indicated only partial

agreement between architects and residents as to what
constitutes a high-quality residential environment. To the
residents, key attributes indicative of overall quality
included general upkeep, noise level and degree of
separation from neighbors.

1212. Marans, R. W., & Rodgers, W. Toward an understanding
       of community satisfaction. In A. H. Hawley & V.
       P. Rock (Eds.), METROPOLITAN AMERICA IN
       CONTEMPORARY PERSPECTIVE. New York: John Wiley &
       Sons, 1975. Pp. 299-352.

   A conceptual model of environmental satisfaction is
presented and then tested. Assessments of particular
environmental attributes, and demographic characteristics as
mediated by these assessments, were found to be related to
community, neighborhood, and housing satisfaction.

1213. Menchik, M. Residential environmental preferences
       and choice: Empirically validating preference
       measures. Environment and Planning, 1972, 4,
       445-458.

   Develops statistical procedures for validation of
preference measures by relating them to attributes of
respondents' current residential environments. Preferences
are viewed as involving trade-offs among accessibility,
characteristics of house and lot, quality of natural
environment, and characteristics of non-natural environment.

1214. Michelson, W. Potential candidates for the
       designer's paradise: A social analysis from a
       nationwide survey. Social Forces, 1967, 46,
       190-196.

   Very little support was found for the assumptions held
by many designers that people prefer multiple dwellings,
limited private open space, central city locations, and mass
transit.

1215. Michelson, W., Belgue, D., & Steward, J. Intentions
       and expectations in differential residential
       satisfaction. Journal of Marriage and the Family,
       1973, 25, 189-198.

     Interviews were conducted with over 700 Toronto families
after they had selected a new dwelling but prior to their
actual move. Husbands and wives agreed on physical
environment factors which influenced their housing choice
but otherwise had somewhat different criteria for selection.
Apartment and home dwellers cited different physical
features as influencing their decisions.

1216. Moriarty, B. M. Socioeconomic status and
       residential locational choice. Environment and
       Behavior, 1974, 6, 448-469.

     Analysis of residential choice and housing distribution
patterns in one metropolitan area indicate that the
locational attributes and preference structure postulated by
the social choice hypothesis are more significant than
economic competition variables in explaining residential
segregation.

1217. Newman, S. J., & Duncan, G. J. Residential problems,
       dissatisfaction, and mobility. Journal of the
       American Planning Association, 1979, 45, 154-166.

     Analyses of longitudinal data for a representative
national sample of families indicated some links between
housing and neighborhood problems and expressions of
discontent, but little strong or direct effect on actual
mobility.

1218. Newman, S. Perceptions of building heights: An
       approach to research and some preliminary
       findings. In #1078 (Conway), pp. 182-193.

Secondary analysis of quality of life data indicates that when factors such as life cycle, location, evaluation of costs, etc. are taken into account, there is no negative impact of high-rise dwelling upon housing satisfaction.

1219. Onibokun, A. G. Evaluating consumers' satisfaction
       with housing: An application of a systems
       approach. Journal of the American Institute of
       Planners, 1974, 40, 189-200.

Relative satisfaction of public housing tenants was assessed with respect to the dwelling unit, surrounding environment, and management. While tenants were generally well satisfied, levels of satisfaction with the environmental sub-system were somewhat lower than for other domains.

1220. Onibokun, A. G. Social system correlates of
       residential satisfaction. Environment and
       Behavior, 1976, 8, 323-344.

Satisfaction with public housing was found to be associated with family size, marital and socio-economic status, tenure of residence, previous residence, and self-perceptions of social class.

1221. Onibokun, A. G. Environmental issues in housing
       habitability. Environment and Planning, 1973, 5,
       461-476.

Interview responses of tenants in 15 public housing sites were factor analyzed into six dimensions of housing habitability. The most pervasive habitability problems were found in the highest density developments.

1222. Peterson, G. L. Measuring visual preferences of
       residential neighborhoods. Ekistics, 1967, 23,
       169-173.

Twenty-three photographs of residential areas were rated
in terms of eight environmental variables as well as
preference. Analyses indicated the variables "beauty" and
"safety" to be nearly colinear with preference.

1223.   Rent, G. S., & Rent, C. S. Low-income housing:
        Factors related to residential satisfaction.
        Environment and Behavior, 1978, 10, 459-488.

A high degree of satisfaction was found for both housing
unit and neighborhood, with housing satisfaction related to
ownership, single family dwelling, positive sentiments
toward neighbors, and a positive life orientation.

1224.   Wekerle, G. Residential choice and housing
        satisfaction in a singles high-rise complex.   In
        #1078 (Conway), pp. 230-239.

It is argued that the high-rise is not a negative living
environment. As housing for young adults, most of whom see
such settings as appropriate for a certain phase of their
lives, high-rise environments can work very well.

1225.   Western, J. S., Weldon, P. D., & Huang, T. T.
        Housing and satisfaction with environment in
        Singapore. Journal of the American Institute of
        Planners, 1974, 40, 201-208.

A survey of residents of Singapore living in both
tenements and government housing. The form of housing per
se had little effect on residents' satisfaction with their
environment. It is suggested that more specific aspects of
housing (i.e., both physical features and socio-demographic
measures) may predict satisfaction differentially across
various living environments.

1226. Williams, J. A. The multi-family housing solution
       and housing type preferences. Social Science
       Quarterly, 1971, 52, 543-559.

   Among a sample of low-income families in the Southwest,
single family dwellings were by far the preferred housing
type. Nearly half of the sample, however, would accept
multi-family housing if amenities such as privacy, outdoor
space, and an option to purchase were provided.

1227. Wilson, R. L. Liveability of the city: Attitudes
       and urban development. In F. S. Chapin & S. Weiss
       (Eds.), URBAN GROWTH DYNAMICS. New York: John
       Wiley, 1962. Pp. 359-399.

   An in-depth investigation of residential satisfaction in
two medium-sized North Carolina cities. Respondents' ideal
neighborhood was characterized as being country-like and
close to nature.

1228. Yeh, S. H. K. Homes for the people. Ekistics, 1974,
       38, 35-41.

   A survey of over 7,000 residents of government housing
indicated satisfaction with housing to be a consequence of
social relations with neighbors, size of the flat, and
perception of change in life situation. The impact of the
physical setting increased when relations with neighbors
were viewed as unimportant.

1229. Zehner, R. B. Neighborhood and community
       satisfaction: A report on new towns and less
       planned suburbs. In #14 (Wohlwill & Carson), pp.
       169-183.

   Residents of new towns rated their communities and
microneighborhoods more highly than did residents of less
planned suburbs. While accessibility to work, shopping, and

other facilities is important in community ratings, it had little impact on neihborhood evaluation.

See also: citations 620, 630, 1402, 1405, 1406, 1421.

12. THE URBAN ENVIRONMENT

The environment-behavior literature related to the urban environment is in many ways as complex and diverse as the city itself. While traditionally the province of urban sociology, the urban crises of the past two decades have prompted considerable research into the nature and consequences of city life from a social psychological perspective as well. Some of the most recent work in this area has endeavored to apply and integrate environment-behavior theory and research on such concepts as density, crowding and sensory overload with more traditional views of urban life.

These diverse disciplinary, theoretical and methodological approaches to the study of the urban environment, and the still embryonic efforts at integration would seem to defy any simple categorization. Thus the topical organization of the materials in this chapter is intended primarily to provide some measure of convenience to the reader and to clarify relationships to various of the more basic psychological processes considered in Section II. This chapter, therefore is organized in four sections.

a. The Urban Environment. This section includes basic texts, readers and bibliographies which deal with the urban environment in a general fashion. Additional materials in this section reflect the on-going controversy regarding the presumably pathogenic qualities of the urban environment; urban-rural comparisons of mental health, concepts of sensory and social overload, and effects of environmental stressors endemic to the city all fall within this domain. Much of this material on the effects of urban experience relates directly to the topics of stimulation, stress, and socio-spatial processes reviewed in Chapters 4 and 5.

b. Behavior in Urban Spaces. Studies in this section
focus, typically in more individual than aggregate fashion,
on a range of behaviors within various urban settings.
Included here are studies of pedestrian pace, behavior in
public places such as parks or subways, and helping behavior
on city streets.

c. The Neighborhood and Social Relationships. The
viability of the concept of "neighborhood" and its
definition in physical and/or social terms has been an
enduring theme in much sociological and planning research.
The effects of neighborhood and street configuration and
urban versus suburban location on patterns of neighboring
and neighborhood attachment are among the issues considered
in this section.

d. Urban Perception and Cognition. Much of the research
in the area of environmental perception and cognition is
rooted within the design and planning professions, with the
majority of these studies conducted in urban settings. Thus
this section includes some of the substantial literature on
the ways in which people perceive, cognize, and evaluate the
city. Much related material may be found in Chapter 4.

a. The Urban Environment

1230. Baldassare, M. Crowding and human behavior: Are
        cities behavioral sinks? Monticello, Ill.:
        Council of Planning Librarians, 1974. (Exchange
        Bibliographies 631). 15 pp.

     Includes a brief review paper and non-annotated
bibliography focusing on theoretical perspectives on urban
crowding as well as field and experimental studies of the
effects of density.

1231. Bell, G., Randall, E., & Roeder, J. E. URBAN
        ENVIRONMENTS AND HUMAN BEHAVIOR. Stroudsburg,
        Pa.: Dowden, Hutchinson & Ross, 1973. xvi, 271
        pp.

A literature review and annotated bibliography, still
quite useful, including many classic references. Sections
are devoted to both design and social science approaches to
the urban environment. A final section focuses on subsets
of the urban environment, ranging in scale from rooms to
city centers.

1232. Brown, L. An annotated bibliography of the
       literature on livability, with an introduction and
       an analysis of the literature. Monticello, Ill.:
       Council of Planning Librarians, 1975. (Exchange
       Bibliographies #853). 61 pp.

Includes an extensive bibliography (partially annotated)
and a review of the literature dealing with such facets of
livability as attitudes toward the city, urban form,
density, location and journey to work, and measurement of
livability.

1233. Faris, R. E. L., & Dunham, H. W. MENTAL DISORDERS IN
       URBAN AREAS. Chicago: University of Chicago
       Press, 1939. 270 pp.

Classic ecological study documenting rates of mental
illness in different spatial areas of the city, illustrating
a systematic decline from the center of the city to the
periphery.

1234. Fischer, C. S. THE URBAN EXPERIENCE. New York:
       Harcourt, Brace, Jovanovich, 1976. 309 pp.

The social and psychological consequences of urban life
are analyzed in terms of the three primary sociological
theories of urbanism. Chapter 3 focuses on effects of the
physical context of the city.

1235. Helmer, J., & Eddington, N. A. (Eds.), URBANMAN.
       New York: The Free Press, 1973. xii, 274 pp.

A collection of papers, some original and others well-
known, examining the ways in which the quality of city life
affects the individual. Particular attention is paid to
commonplace activities such as walking, driving or waiting
in lines, and to problems of apathy, vandalism, and
psychopathology.

1236. Kalt, N. C., & Zalkind (Eds.), S. S. URBAN PROBLEMS:
       PSYCHOLOGICAL INQUIRIES. New York: Oxford
       University Press, 1976. 543 pp.

A reader dealing with a variety of urban issues. Of
particular relevance are Chapters 3 and 8 focusing on
housing and effects of the urban environment.

1237. Karp, D. A., Stone, G. P., & Yoels, W. C. BEING
       URBAN: A SOCIAL PSYCHOLOGICAL VIEW OF CITY LIFE.
       Lexington, Mass.: D.C. Heath and Co., 1977. 242
       pp.

Attention is focused on some of the social psychological
questions typically neglected in treatments of urban life.
Issues considered include patterns of urban interaction and
tolerance, and "symbolic" as opposed to spatial definitions
of the city.

1238. Lynch, K. GROWING UP IN CITIES. Cambridge, Mass.:
       MIT Press, 1977. 177 pp.

Small groups of teen-age children in cities in
Argentina, Australia, Mexico and Poland were asked to
describe and map their urban environment. Excerpts from the
four national reports are included along with a comparative
summary.

1239. Michelson, W. MAN AND HIS URBAN ENVIRONMENT: A
       SOCIOLOGICAL APPROACH (2nd edition). Reading,
       Mass.: Addison-Wesley, 1976. xiii, 273 pp.

   Building upon a model of experiential congruence,
various concepts from personality, cultural and social
systems are related to relevant aspects of the environmental
system, and studies within each domain are reviewed. A new
concluding chapter has been added to the second edition.

1240. Rapoport, A. HUMAN ASPECTS OF URBAN FORM. Oxford:
       Pergamon Press, 1977. 438 pp.

   An application of man-environment studies to issues of
urban form and design. Covers perspectives such as
environmental perception and cognition, ethology, activity
systems and networks. The volume includes a bibliography of
over 1,100 citations.

1241. Sadalla, E. K., & Stea, D. (Eds.) Approaches to a
       psychology of urban life. Environment and
       Behavior. 1978, 10, 139-291.

   A special issue including contributions on the nature of
urban environmental perception, the relationships between
density and psychopathology, city size and human behavior,
the concept of place-identity, and the approach of
ecological psychologists to the urban environment.

1242. Appleyard, D., & Lintell, M. The environmental
       quality of city streets: The residents' viewpoint.
       Journal of the American Institute of Planners,
       1972, 38, 84-101.

   Traffic levels on three city streets were found to be
inversely correlated with extent of home territory, social

interaction and safety, and directly related to problems of
noise and pollution.

1243. Baldassare, M., & Fischer, C. S. The relevance of
          crowding experiments to urban studies. In #13
          (Stokols), pp   273-285.

It is argued that most studies of crowding fail to
consider and build upon any of several theoretical
perspectives regarding the nature of urban life. A number
of researchable questions, based upon each of these
perspectives, are presented.

1244. Bechtel, R. B. A behavioral comparison of urban and
          small town environments. In #35 (EDRA-2), pp.
          347-353.

Adaptation of Barker's behavior setting methodology to
an urban area revealed that while city block residents have
more behavior settings available to them than do small town
residents they do not participate as actively in controlling
them. It is suggested that urban areas should be designed
so as to create more socially demanding settings.

1245. Broady, M. The sociology of the urban environment.
          Ekistics, 1970, 29, 184-190.

The assumption that a city's social and economic life is
a consequence of its physical form is questioned.
Organization theory is suggested as an alternative approach
to the study of urban environments and is applied to the
analysis of a British New Town.

1246. Carnahan, D., Cove, W., & Galle, O. R. Urbanization,
          population density, and overcrowding: Trends in
          the quality of life in urban America. Social
          Forces, 1974, 53, 62-72.

Analyses of Housing Census data for 1940-1970 indicate a
decline in housing density for the nation as a whole, and
that households in central cities and SMSA'S are no more
crowded than the national average. Thus it is suggestd that
this form of density does not account for increases in
social pathologies in urban areas. It is also proposed that
the framework of much density research, both in detail and
in total, be further examined.

1247.   Carp, F. M. Life-style and location within the city.
        The Gerontologist, 1975, 15, 27-33.

        Centrality of residence of older persons within the city
was associated with more favorable life-style, as defined in
terms of active, autonomous and satisfying use of time,
space, and social networks. Data suggest that the obvious
drawbacks of the center city as a residence for the elderly
are outweighed by its advantages.

1248.   Carp, F. M., Zawadski, R. T., & Shokrkon, H.
        Dimensions of urban environmental quality.
        Environment and Behavior, 1976, 8, 239-264.

        Analysis of ratings of 100 questionnaire items
descriptive of environmental quality yielded 20 factors
which were interpreted in terms of six clusters of
dimensions dealing with noise, aesthetics, neighbors,
safety, mobility, and annoyances.

1249.   Dohrenwend, B. S., & Dohrenwend, B. P. Psychiatric
        disorders in urban settings. In S. Arieti et al.
        (Eds.), AMERICAN HANDBOOK OF PSYCHIATRY. (2nd
        ed.). New York: Basic Books, 1974. Pp. 424-447.

        A review of a substantial number of studies comparing
urban and rural mental health, indicating that psychoses
were more prevalent in rural areas, while both neuroses and
personality disorders were more prevalent in urban areas.

1250. Fischer, C. S. A research note on urbanism and
       tolerance. American Journal of Sociology, 1971,
       76, 847-856.

       It was found that there is no direct relationship
between urbanism and increased inter-personal tolerance.
The relatively greater tolerance in cities is more a
function of social characteristics of inhabitants than city
size per se.

1251. Fischer, C. S. The metropolitan experience. In A.H.
       Hawley & V.P. Rock (Eds.), METROPOLITAN AMERICA IN
       CONTEMPORARY PERSPECTIVE. New York: Wiley, 1975.
       Pp. 201-234.

       An analysis of the effects of urban life - and
particularly the size of place - on individual experience.
It is argued that the few differences which do emerge are
largely mediated by social class. The city, however, may be
distinctive in terms of the less tangible attitudes, values
and potential experiences which it may provide or permit.

1252. Fischer, C. S., Baldassare, M., & Ofshe, R. J.
       Crowding studies and urban life: A critical
       review. Journal of the American Institute of
       Planners, 1975, 41, 406-418.

       The existing research on the effects of population
density is reviewed and its often tenuous links to urban
studies considered. Several theoretical frameworks to guide
deductive research are presented.

1253. Gifford, R., & Peacock, J. Crowding: More fearsome
       than crime-provoking? Comparison of an Asian city
       and a North American city. Psychologia, 1979, 22,
       79-83.

Former residents of Toronto, where the crime rate is four times as great but the density is one quarter that of Hong Kong, reported more personal memories of crime but had felt significantly safer than had former residents of Hong Kong. It is suggested that crowding may be associated more with the fear of crime than with its actual occurrence.

1254. Gillis, A. R. Population density and social pathology: The case of building type, social allowance, and juvenile delinquency. Social Forces, 1974, 53, 306-314.

Internal density (persons per room), external density (persons per residential acre) and building type (multiple versus single dwelling) were related to rates of public assistance and juvenile delinquency through ex post facto analyses of 30 city census tracts. After controlling for effects of income and national origin, only building type was significantly related to both dependent variables. A path model is developed and issues of causality are considered.

1255. Gump, P. V., & Adelberg, B. Urbanism from the perspective of ecological psychologists. In #1241 (Sadalla & Stea), pp. 171-191.

It is suggested that the size of a community is more meaningfully defined in terms of number of behavior settings than population. Findings regarding activities of children in small towns, a small city, and a neighborhood of a large metropolis are reviewed and compared. Finally a descriptive model for urbanism is proposed, incorporating the ecological habitat, behavior and experience of inhabitants, and intraperson changes in response to habitat.

1256. Kirmeyer, S. L. Urban density and pathology. In #1241 (Sadalla & Stea), pp. 247-269.

A review of the research on the association between
areal and in-dwelling densities and health and social
pathologies. Limitations of aggregate, correlational
studies and reliance on archival measures of pathology are
considered. While aggregate urban density appears to have
little effect on pathology independent of socioeconomic
variables, there may be negative impacts on quality of life.

1257. Korte, C. Helpfulness in the urban environment. In
        #12 (Baum et al.), pp. 85-109.

A review of research on the nature and incidence of
mutual aid in urban environments. Attention is given to
both environmental and cultural limitations.

1258. Merrens, M. Nonemergency helping behavior in various
        sized communities. Journal of Social Psychology,
        1973, 90, 327-328.

In six of eight cases residents in midwestern cities
and town provided significantly greater helping behavior
than pedestrians in New York City.

1259. Michelson, W. An empirical analysis of urban
        environmental preferences. Journal of the
        American Institute of Planners, 1966, 32, 355-360.

A combination of interviews, rankings of photographs and
map drawing was employed to study the preferences and ideal
urban environments of a sample of city residents, in terms
of preferred types, scale and spatial configurations of
housing and neighborhoods.

1260. Milgram, S. The experience of living in cities.
        Science, 1970, 167, 1461-1468.

Used the concepts of social stimulus overload and
adaptation to such overload in analyzing differences in
levels of trust, social responsibility, and helping behavior
reported in studies of urban and town dwellers.

1261.   Roberts, C. Stressful experiences in urban places:
        Some implications for design. In #41 (EDRA-8),
        pp. 182-194.

Stress situations reported by residents of Manhattan
were characterized as relating to the social and physical
environments, crime, and basic living problems. By far the
most stress was experienced in public spaces and
particularly in those used to get from one place to another.
Differences related to sex and length of residency of
respondent emerged in the areas of crime and basic living
problems.

1262.   Sadalla, E. K. Population size, structural
        differentiation, and human behavior. In #1241
        (Sadalla & Stea), pp. 271-291.

A review article suggesting that the size of an urban
area, along with concomitant social differentiation, is
related to such variables as anonymity, deindividuation, and
other facets of personality development and social
interaction.

1263.   Schmitt, R. C. Density, health, and social
        disorganization. Journal of the American
        Institute of Planners, 1966, 32, 38-40.

Correlational analyses indicated that of five measures
of density and residential overcrowding, population per net
acre had the strongest association with mortality, morbidity
and health rates for 42 census tracts. It is suggested that
high density may be associated with decrements in health and
social organization.

1264. Schmitt, R. C., Zane, L. Y., & Nishi, S. Density,
       health and social disorganization revisited.
       Journal of the American Institute of Planners,
       1978, 44, 209-211.

       Analysis of up-dated data for the same census tracts
studied previously (item #1263) revealed a decline in
correlations between resident population densities and
various measures of health and social disorganization.

1265. Singer, J. E., Lundberg, U. & Frankenhaeuser, M.
            Stress on the train: A study of urban commuting.
            In #12 (Baum et al.), pp. 41-56.

       Stress arising from daily commuting to work was assessed
through both passengers' qualitative assessment of the trip
and psychophysiological measures. Stress was found to vary
more with social and ecological circumstances (e.g. greater
crowding, lack of choice in seat selection) than with length
or duration of train trip.

1266. Srole, L. Urbanization and mental health: Some
            reformulations. American Scientist, 1972, 60,
            576-583.

       A review of thinking and evidence on mental-health
related differences between urban and rural residents, and a
reformulation of the alleged role of urban life in promoting
mental illness and social pathology. The role of selective
factors attracting certain types of individuals to those
environments optimally fitting their personality traits is
stressed.

1267. Webb S. D. Mental health in rural and urban
            environments. Ekistics, 1978, 45, 37-42.

       A nationwide sample in New Zealand revealed only a very
limited relationship between the prevalence of mental

illness and urban versus rural residence. The marginal
differences suggested increased impairment for small town
and small town fringe inhabitants.

See also: citations 1, 12, 28, 158, 273, 287, 295, 302, 305,
347, 398, 417, 423, 428, 445, 619, 861.

b. Behavior in Urban Spaces

1268. Chapin, F. S. HUMAN ACTIVITY PATTERNS IN THE CITY.
       New York: Wiley, 1974. xii, 272 pp.

A broad range of activities engaged in by urbanites was
classified into 12 main behavioral catagories associated
with employment, family and household activities,
socializing, recreation and organizational participation.
Activity patterns were found to vary with demographic
variables and temporal patterns.

1269. Whyte, W. H. THE SOCIAL LIFE OF SMALL URBAN SPACES.
       Washington, D. C.: The Conservation Foundation,
       1980. 125 pp.

A review of Whyte's observational research on urban
streets and spaces. People were found to congregate in
plazas and other spaces which provided sun exposure, trees,
water, food, and places to sit and watch other people.
Appendices review Whyte's time-lapse filming technique as
well as the New York City zoning ammendments which emerged
from his research.

1270. Becker, F. D. A class-conscious evaluation: Going
       back to Sacramento's pedestrian mall. Landscape
       Architecture Quarterly, 1973, 64, 448-457.

An evaluation of a downtown urban mall three years after
its completion. Results suggested that merchants viewed the
mall as a mechanism for economic revitalization, middle-
class users viewed it as a place to shop, and young people
viewed it as a social/entertainment setting.

1271. Bornstein, M. H., & Bornstein, H. G. The pace of
       life. Nature, 1976, 259, 557-559.

Systematic observations in 15 cities and towns in
Europe, Asia and North America indicate that the pace of
pedestrian movement varies directly with size of local
population, regardless of the cultural setting.

1272. Brower, S. Streetfront and sidewalk. Landscape
       Architecture Quarterly, 1973, 63, 364-369.

Summer-long observations of a predominantly black,
lower-income area of Baltimore revealed a striking tendency
to utilize front steps and sidewalks rather than rear yards
or playgrounds. Proposed explanations are based upon
interviews with residents and are related to patterns of
social life, security considerations and parental
supervision.

1273. Brower, S., & Williamson, P. Outdoor recreation as a
       function of the urban housing environment.
       Environment and Behavior, 1974, 6, 295-345.

A study of utilization of outdoor space was conducted in
three Baltimore neighborhoods. In contrast to the two
inner-city neighborhoods where substantial leisure time was
spent on the streetfront, most sitting out and playing in
the middle-income neighborhood took place in back yards and
parks.

1274.   Cohen, H., Moss, S., & Zube, E. Pedestrians and wind
        in the urban environment. In #43 (EDRA-10), pp.
        71-82.

     Pedestrian behavior was observed and analyzed adjacent
to two buildings in downtown Boston. On test days (i.e.
ambient wind speed greater than 20 m.p.h.) pedestrian speed
increased, pedestrian density and various other forms of
pedestrian behavior decreased, and circulation patterns were
less varied and more directional.

1275.   Dornbusch, D. M., & Gelb, P. M. High-rise impacts on
        the use of parks and plazas. In #1078 (Conway),
        pp. 112-130.

     A survey of users of an urban park indicated that the
presence of high-rise buildings tended to limit both
visibility and utilization of the park; more users,
travelling greater distances, came from those areas most
open to the park.

1276.   Effran, M. G., & Baran, G. S. Privacy regulation in
        public bathrooms: Verbal and nonverbal behavior.
        In #1468 (Esser & Greenbie), pp. 215-221.

     Within a public bathroom, privacy regulation mechanisms
were found to be operative for men at urinals but not at
sinks. Analogies are drawn to behavior in urban public
places such as subways and elevators.

1277.   Gelb, P. M. High-rise impact on city and
        neighborhood livability. In #1078 (Conway), pp.
        131-145.

     Observations on pairs of residential and commercial city
blocks revealed substantially less street life

(conversations, playing children, etc.) on high-rise as
contrasted with low-rise blocks.

1278. Kamman, R., Thomson, R., & Erwin, R. Unhelpful
        behavior in the street: City size or immediate
        pedestrian density. Environment and Behavior,
        1979, 11, 245-250.

City size has been found to be correlated with immediate
pedestrian density, which in turn has been a reliable
predictor of nonhelping behavior. This pattern is confirmed
in this study. It is proposed that the diffusion-of-
responsibility model offered by Latane and Darley gives a
good account of the data on helping behavior in the street.

1279. Korte, C., Ypma, I., & Toppen, A. Helpfulness in
        Dutch society as a function of urbanization and
        environmental input level. Journal of Personality
        and Social Psychology, 1975, 32, 996-1003.

A comparison of three different measures of helping
behavior in four Amsterdam neighborhoods; a preliminary
survey of local residents had characterized two of these
areas as very helpful and friendly and the other two as not
very helpful or friendly. Contrary to expectations, no
differences in helpfulness were found.

1280. Levine, J., Vinson, A., & Wood, D. Subway behavior.
        In A. Birenbaum & E. Sagarin (Eds.), PEOPLE IN
        PLACES: THE SOCIOLOGY OF THE FAMILIAR. New York:
        Praeger, 1973. Pp. 208-216.

Subway riders were found to choose seats which maximized
their distance from fellow travellers. Visual attention was
limited to visual props such as newspapers or to looking out
of windows.

1281. Love, R. L. The fountains of urban life. Urban Life and Culture, 1973, 2, 161-208.

The watching of other people was found to be a central reason for visiting a Portland, Oregon park which attracted a wide diversity of persons. Within the park itself there was both an ecological segregation of persons and activities and a high degree of tolerance of diversity.

1282. Lowin, A., et al. The pace of life and sensitivity to time in urban and rural settings. Journal of Social Psychology, 1971, 83, 247-253.

Field studies carried out in six metropolitan areas and in sets of small towns provide only partial validation for the stereotypical view of activities as being faster paced in urban areas. An anticipated greater time-accuracy of urban dwellers could not be substantiated.

1283. Mann, L. Learning to live with lines. In #1235 (Helmer & Eddington), pp. 42-61.

An analysis of the common urban experience of waiting in queues. Attention is focused on the self-policing nature of queues and problems of queue jumping, as well as the more positive social aspects of waiting in lines.

1284. Milgram, S. & Sabini, J. On maintaining urban norms: A field experiment in the subway. In #12 (Baum et al.), pp. 31-40.

This study probed norms and patterns of social control within the subway environment by simply having experimenters ask passengers if they would give them their seats. Contrary to expectations, a majority of requests resulted in the offer of a seat. The power of such norms was reflected in the difficulties experienced by a number of experimenters in carrying out this request.

1285. Newman, J., & McCauley, C. Eye contact with
       strangers in city, suburb, and small town.
       Environment and Behavior, 1977, 9, 547-558.

     Eye contact among strangers was found to be primarily a
function of locale, as opposed to other variables such as
sex of experimenter. It is suggested that reduced levels of
social interaction in cities represent an adaptation to
overload rather than an expression of pathology.

1286. Rushton, J. P. Urban density and altruism: helping
       strangers in a Canadian city, suburb, and small
       town. Psychological Reports, 1978, 43, 987-990.

     Helpfulness to strangers (in terms of readiness to give
the time, directions, change of a quarter, and own name) was
found to be inversely related to the density of the area in
which the request was made.

1287. Share, L. B. Giannini Plaza and Transamerica Park:
       Effects of their physical characteristics on
       users' perceptions and experiences. In #42
       (EDRA-9), pp. 127-139.

     A comparative study of two physically proximate but
spatially different open spaces in downtown San Francisco.
Interviews with users indicated that these spaces provided
relief and contrast from both work place and the larger
urban environment; natural elements and exposure to sunshine
were particularly important in meeting users' desires.

1288. Van Vliet, W. User evaluation of concentrated
       facilities. Ekistics, 1978, 45, 441-443.

     Residents' evaluations and utilization of the
multifunctional commercial core of a Dutch new town were
differentiated by length of residence, stage in the life
cycle, and proximity of home to the center.

1289.   Whyte, W. H., with Bemiss, M.   New York and Tokyo: A
        study in crowding.   Real Estate Issues, 1979, 4,
        1-17.

        A comparison of observational studies of pedestrian
behavior carried out in center city areas of Tokyo and New
York.   While there were some differences in numbers of
people walking in groups and in group size, the most
preferred places in both cities were those where large
numbers of people were clustered in relatively limited
space.   It is suggested that the actual experiences of many
urbanites run counter to the view of the city as a stressful
setting of sensory and social overload.

1290.   Wolff, M.   Notes on the behavior of pedestrians.   In
        A. Birenbaum & E. Sagarin (Eds.), PEOPLE IN
        PLACES: THE SOCIOLOGY OF THE FAMILIAR.   New York:
        Praeger, 1973.   Pp. 35-48.

        A study of the distance at which pedestrians "yield" to
an experimenter approaching on a collision path along a city
street.   Yield distances were found to vary with density of
the situation and sex of the experimenter.

1291.   Zimbardo, P. G.   A field experiment in auto shaping.
        In C. Ward (Ed.), VANDALISM.   New York: Van
        Nostrand Reinhold, 1973.   Pp. 85-90.

        A car abandoned in New York City was stripped of all
movable parts within 24 hours after which it was subjected
to random destruction, typically during daylight hours by
well-dressed individuals.   A comparable car abandoned in
Palo Alto was left untouched.

See also: citations 441, 563

c. The Neighborhood and Social Relationships

1292.   Fischer, C. S., et al. NETWORKS AND PLACES: SOCIAL
    RELATIONS IN THE URBAN SETTING. New York: The
    Free Press, 1977. 229 pp.

An introduction to and exploration of social networks
and network analysis. Part II analyzes the interactions
between social networks and the places people live.

1293.   Gans, H. J. THE URBAN VILLAGERS: GROUP AND CLASS IN
    THE LIFE OF ITALIAN AMERICANS. New York: Free
    Press of Glencoe, 1962. xvi, 367 pp.

Gans' experiences as a participant-observer revealed
that the West End of Boston, while labelled a slum and
levelled through urban renewal, was in many ways a healthy
working class neighborhood with strong local social
relationships and sense of community.

1294.   Gans, H. J. THE LEVITTOWNERS. New York: Pantheon,
    1967. xxix, 474 pp.

An application of Gans' participant-observation approach
to a new suburban community. Findings focus on the origins
of the community, the quality of suburban social life, and
community politics.

1295.   Keller, S. THE URBAN NEIGHBORHOOD. New York: Random
    House, 1968. 201 pp.

A sociological inquiry into factors affecting
neighboring, neighborhood attachment, and satisfaction. The
concluding chapter focuses on implications for planning
practice.

1296. Newman O. COMMUNITY OF INTEREST. Garden City,
       N.Y.: Anchor Press, 1980. 356 pp.

An expansion of concepts previously developed at the
housing scale (item #1086) to deal with the stabilization
and development of viable communities. Drawing upon
examples from several cities, it is proposed that urban
environments be constituted of small community enclaves of
like-minded people brought together by mutual and limited
self-interest; implications for both physical design and
social policy are considered.

1297. Suttles, G. D. THE SOCIAL CONSTRUCTION OF
       COMMUNITIES. Chicago: The University of Chicago
       Press, 1972. 278 pp.

An effort to integrate the traditional concept of the
"natural community" as developed by the early human
ecologists with more recent concepts of territorial
behavior. It is argued that American communities have
become overly compartmentalized, localizing and intensifying
social hostilities; examples are drawn from a redevelopment
area on Chicago's South Side.

1298. Whyte, W. H. THE ORGANIZATION MAN. New York: Simon
       & Schuster, 1956. 420 pp.

Whyte's classic analysis of suburbia includes a number
of observations of social interaction within Park Forest,
Illinois. Patterns of friendship are related to variations
in location, street width and traffic patterns, and
utilization of shared spaces.

1299. Baldassare, M. Residential density, local ties and
       neighborhood attitudes: Are the findings of

microstudies relevant to urban areas?
Sociological Symposium, 1975, 14, 93-102.

Decreased social behavior, observed in the laboratory as
an effect of crowding, was not found among residents of city
neighborhoods of high density.

1300. Baldassare, M. Residential density, household
       crowding, and social networks. In C.S. Fischer,
       et al. (Eds.), NETWORKS AND PLACES: SOCIAL
       RELATIONS IN THE URBAN SETTING. New York: The
       Free Press, 1977. Pp. 101-115.

Secondary analyses of two sets of data revealed little
or no impact of either household or neighborhood density on
male residents' social networks.

1301. Canter, M. H. Life space and the social support
       system of the inner city elderly of New York. The
       Gerontologist, 1975, 15, 23-26.

A cross-cultural survey of older persons living in the
poorest areas of New York City revealed the neighborhood to
be defined in quite local terms (e.g. a maximum radius of 10
blocks), and as providing most essential services.
Children, friends and neighbors all provided important
sources of social interaction and personal support.

1302. Caplow, T., & Forman, R. Neighborhood interaction in
       a homogeneous community. American Sociological
       Review, 1950, 50, 357-366.

Classic sociometric study of a married student housing
community. Geographic and inter-personal aspects of
neighborhood were found to be highly congruent.

1303. Chapin, F. S., & Brail, R. K. Human activity systems
       in the metropolitan United States. Environment
       and Behavior, 1969, 1, 107-130.

    Data regarding obligatory and discretionary activities
were gathered for a national sample of over 1,100 heads of
household and spouses in 43 metropolitan areas. Frequency
and duration of activity measures were more strongly related
to personal than to environmental characteristics.

1304. Fischer, C. S., & Jackson, R. M. Suburbanism and
       localism. In #1292 (Fischer et al.), pp. 117-138.

    Suburbanites' social activities were found to be more
local than were those of city-dwellers. This slightly
greater localism was found to reflect both individual traits
and social and physical attributes of the suburban
environment.

1305. Fried, M. Grieving for a lost home: Psychological
       cost of relocation. In L. Duhl (Ed.), THE URBAN
       CONDITION. New York: Basic Books, 1963. Pp.
       151-171.

    Interviews conducted with 366 residents of Boston's West
End prior to and after their forced relocation due to urban
renewal indicated that their affective responses went far
beyond the researchers' expectations and displayed most of
the characteristics of grief and mourning.

1306. Gans, H. J. Planning and social life: Friendship
       and neighbor relations in suburban communities.
       Journal of the American Institute of Planners,
       1961, 27, 134-140.

    It is argued that while site design can bring people
together inititally, homogeneity is far more instrumental
than propinquity in the creation of strong social
relationships.

1307.   Gerson, K., Stueve, C. A., & Fischer, C. S.
        Attachment to place. In #1292 (Fischer et al.),
        pp. 139-160.

Attachment to one's neighborhood was found to be multidimensional in nature, with differing kinds of behavioral
involvement not necessarily related to one another nor to
affect. Levels of attachment were also related to both
social and environmental variables.

1308.   Kasarda, J. D., & Janowitz, M. Community attachment
        in mass society. American Sociological Review,
        1974, 39, 320-339.

A survey conducted in Great Britain revealed no
differences in number of local friendships between residents
of high-density and low-density wards.

1309.   Koyama, T. Rural-urban comparisons of kinship
        relations in Japan. In R. Hill & R. Konig (Eds.),
        FAMILIES IN EAST AND WEST. Paris: Mouton, 1970.
        Pp. 318-337.

While equal proportions of urban and rural families
indicated that they would turn to personal sources of help
when in need, urban residents indicated less reliance on
neighbors than did rural families.

1310.   Mayo, J. M. Effects of street forms on suburban
        neighboring behavior. Environment and Behavior,
        1979, 11, 375-397.

Differing street configurations (i.e. linear,
curvilinear and cul-de-sac) were significantly less
influential than social variables in predicting neighboring
behavior.

1311. McCauley, C., & Taylor, J. Is there overload of
       acquaintances in the city? Environmental
       Psychology and Nonverbal Behavior, 1976, 1, 41-55.

    On the basis of a telephone survey of 176 residents of a
large city and a small town in Canada, the former were found
to have fewer but equally intimate social contacts as the
latter.

1312. Reiss, A. J. The sociological study of community.
       Rural Sociology, 1959, 24, 118-130.

    Time budgets revealed no differences between city and
rural aggregates in average amount of time spent with kin.
Urban men also allocated their time more heavily in favor of
friends than did rural men.

1313. Tomeh, A. K. Informal group participation and
       residential patterns. American Journal of
       Sociology, 1964, 70, 28-35.

    Interviews with over 800 Detroit area adults revealed
suburban living to be associated with more contacts on the
local level, with a gradual increase in contacts with
friends associated with greater distance from the city
center.

1314. Willmott, P. Housing density and town design. Town
       Planning Review, 1962, 33, 115-128.

    A study of attitudes toward density, privacy and
services held by residents of a British New Town. In
contrast to the town's planners, 42% of respondents felt
densities in the community to be too high.

d. Urban Perception and Cognition

1315. Appleyard, D. PLANNING A PLURALISTIC CITY:
        CONFLICTING REALITIES IN CIUDAD GUYANA.
        Cambridge, Mass.: MIT Press, 1976. 312 pp.

     An account of the planning and development of a new city
in a remote area of Venezuela, focusing on the contrasting
and often conflicting conceptions and values of residents
and planners. Research findings regarding residents'
perception, cognition and evaluation of the city are
reviewed.

1316. Lynch, K. WHAT TIME IS THIS PLACE? Cambridge,
        Mass.: The MIT Press, 1972. 277 pp.

     Lynch expands his previous work on the spatial aspects
of urban perception and cognition to include the temporal
image of the city. The quality of this image of time is
seen as related to both personal well-being and successful
environmental change. The importance of place as a symbol
of time is presented through both case studies and
theoretical discussion.

1317. Hurst, M. I CAME TO THE CITY. Boston: Houghton
        Mifflin, 1975.

     The urban images of two distinct groups, professionals
and welfare recipients, living in different areas of
Vancouver were compared. The welfare group held images of
the city even more localized than those of the
professionals, centering on their neighborhood and a few
community services and facilities.

1318. Appleyard, D. Why buildings are known: A predictive
        tool for architects and planners. Environment and
        Behavior, 1969 1, 131-156.

Measures of the form, visibility and use characteristics
of a large sample of buildings in Ciudad Guyana, Venezuela
were found to be correlated with frequency of appearance on
maps drawn by residents, as well as recall after a trip
through the city.

1319. Bishop, J., & Foulsham, J. Children's images of
          Harwich. In #50 (Kuller), pp. 323-340.

Cognitive maps of the route from home to school were
drawn by over 200 nine year old students. Maps were
analyzed in terms of both their content (i.e. landmarks,
roads, areas) and individual/group differences (i.e. gender,
place of residence, mode of travel).

1320. Carr, S. The city of the mind. In #29 (Proshansky
          et al., 1970), pp. 518-533.

The author proposes that through perceiving and
representing the environment, acting in it, and reviewing
the consequences of these actions we create our own city of
the mind. Proper organization of the city can make tasks in
the city easier to accomplish, increase our scope of
possible actions, and provide an enhanced sense of
competence.

1321. Cox, K. R., McCarthy, J. J., & Nartowicz, F. The
          cognitive organization of the North American city:
          Empirical evidence. Environment and Planning A,
          1979, 11, 327-334.

Reports two studies which explored generic mental
representations of urban areas in terms of associated land
uses and social characteristics of different housing groups.

1322. Dornbusch, D. M., & Gelb, P. M. High-rise visual
       impact. In #1078 (Conway), pp. 101-111.

   A survey of occupants of both commercial and residential
areas in San Francisco indicated that aesthetics represent
the most important criterion of areawide quality for most
respondents and that high-rise developments often run
counter to respondents' aesthetic goals. The greatest
objections were voiced to high-rise developments in
residential areas.

1323. Firey, W. Sentiment and symbolism as ecological
       variables. American Sociological Review, 1945,
       10, 140-148.

   Counter to the traditional ecological position, it is
argued that physical features of the city cannot, by
themselves, explain urban behavior and social organization.
Boston's Beacon Hill is presented as one example of the
importance of mutual symbolic importance placed on specific
areas.

1324. Goodchild, B. Class differences in environmental
       perception. Urban Studies, 1974, 11, 157-169.

   Interviews were conducted with matched samples of
middle-class and working class residents of a British town
regarding their environmental perceptions and preferences.
Middle-class residents were found to be somewhat more
concerned with aesthetic factors, to make more mention of
townscape features, and to draw more extensive sketch maps
with more features and greater conceptual clarity.

1325. Haney, W. G., & Knowles, E. S. Perception of
       neighborhoods by city and suburban residents.
       Human Ecology, 1978, 6, 201-214.

The imageability of neighborhoods in a medium-sized
metropolitan area was found to be both high and consensual.
Suburban neighborhoods, however, were seen as larger and
less negative than those in the inner-city. Results are
seen as supporting and extending the local community model
of neighborhood.

1326. Hansvick, C. L. Comparing urban images: A
         multivariate approach. In #42 (EDRA-9), pp.
         109-126.

Based upon responses to 25 bipolar descriptive
adjectives and 11 environmental features, substantial
differences were found between the images of 10 Canadian
cities. More objective measures of environmental quality
did not directly correspond with respondents' urban images.

1327. Harrison, J. D., & Howard, W. A. The role of meaning
         in the urban image. Environment and Behavior,
         1972, 4, 389-411.

Elements included in 44 subjects' sketch maps of a
Denver suburb were analyzed in terms of four components of
imageability. Subjects' descriptions of these elements
focused more on meaning and location than appearance or
association. Findings are seen as reflecting both the
importance of meaning and the physical characteristics of
the setting studied.

1328. Heinemeyer, W. F. The urban core as a center of
         attraction. In URBAN CORE AND INNER CITY.
         Leiden: E. J. Brill, 1967. Pp. 82-99.

An analysis of the urban core of Amsterdam indicated
that those places perceived as particularly attractive
incorporated a range of functions which served to draw
together a diverse public and create a "forum quality".

1329. Herzog, T. R., Kaplan, S., & Kaplan, R. The
       prediction of preference for familiar urban
       places. Environment and Behavior, 1976, 8,
       627-645.

Nonmetric factor analysis of preference ratings for 70
urban scenes yielded five urban dimensions: Cultural,
Contemporary, Commercial, Entertainment, and Campus. While
both familiarity and complexity were factors in accounting
for preference scores, the patterns of relationships were
quite different across the five urban dimensions.
Additional study conditions, where subjects were presented
verbal labels rather than slides yielded comparable
preference ratings.

1330. Ittelson, W. H. Environmental perception and urban
       experience. In #1241 (Sadalla & Stea), pp.
       193-213.

Dimensions of urban experience are analyzed in terms of
environment as external object, representation of self,
embodiment of value, and arena for action.

1331. Klein, H. J. The definition of the town-center in
       the image of its citizens. In URBAN CORE AND
       INNER CITY. Leiden: E. F. Brill, 1967. Pp.
       286-306.

The consensual town center of Karlsruhe, West Germany
was defined through interviews with over 1,000 residents,
each of whom made judgments regarding which places and roads
were part of the central area. Location of residence, sex,
and socioeconomic status influenced the images held by
respondents.

1332. Kreimer, A. Building the imagery of San Francisco:
       An analysis of controversy over high-rise
       development, 1970-71. In #37 (EDRA-4), vol. 2,
       pp. 221-231.

An exploration of the extent to which perception and
evaluation of the urban environment is shaped by the
communication media. Attention is focused on the
controversy of over high-rise construction in San Francisco
and the role of local newspapers in attributing
environmental characteristics to and creating a mythology
about the city.

1333.   Maurer, R., & Baxter, J. C.   Images of the
            neighborhood and city among Black, Anglo-, and
            Mexican-American Children. Environment and
            Behavior, 1972, 4, 351-388.

Ninety-one children living in a low-income area of
Houston were asked to draw maps of their neighborhood and of
the city. Racial differences were observed in extent of
neighborhood, size and location of child's home, inclusion
of natural versus non-natural elements, and willingness to
draw the city map. Several interpretations of the findings,
focusing on life-style differences, are considered.

1334.   Milgram, S., et al   A psychological map of New York
            City. American Scientist, 1972, 60, 194-200.

Two hundred residents of the five boroughs of New York
City were presented with 152 slides of the city taken from
systematically selected sites. Subjects were asked to
locate each scene by borough, neighborhood, and street.
Irrespective of subjects' borough of residence, slides of
Manhattan were far more likely to be correctly located in
all three tasks.

1335.   Nasar, J. L.   The evaluative image of a city. In #43
            (EDRA-10), pp. 38-45.

Both residents and visitors to a southern city disliked
the central business district and certain highly
commercialized traffic arteries; both groups liked an upper

income residential neighborhood and several exurban areas.
The most frequently cited reasons for preferences included
landscaping, views, modernity, organization, and visible
countryside. Negative evaluations were associated with
signs, dilapidation, industry, and disorganized and crowded
commercial areas.

1336.   Orleans, P., & Schmidt, S.   Mapping the city:
         Environmental cognition of urban residents.   In
         #36 (EDRA-3), vol. 1, sec. 1.4, pp. 1-9.

        A study of husbands' and wives' differing perceptions of
their residential environment as a consequence of
differences in daily activities and extent of local
involvement. Results of a map-drawing task revealed
differences in inclusion of local landmarks, and utilization
of domocentric as opposed to coordinate reference systems.

1337.   Pocock, D. C.   City of the mind: A review of mental
         maps of urban areas.   Scottish Geographical
         Magazine, 1972, 88, 116-124.

        The environment as perceived is seen as central to both
the interpretation of spatial behavior and successful
environmental planning. The non-veridical and ego-centered
nature of cognitive maps is illustrated with examples drawn
from the city of Dundee; particular attention is given to
size and distance distortions.

1338.   Rozelle, R., & Baxter, J.   Meaning and value in
         conceptualizing the environment.   Journal of the
         American Institute of Planners, 1972, 38, 116-122.

        While the elements of Houston which respondents were
able to image and remember were primarily physical features
such as buildings and roads, elements characterized as
important were primarily social and cultural in nature.

1339. Sims, B., & Khan, S. Visual exposure, form,
        uniqueness, social and functional significance
        Toward a predictive model of image formation.
        Man-Environment Systems, 1977, 7, 321-331.

    Building upon earlier work in urban environmental
cognition, recall of landmarks and paths was related to
several aspects of their exposure and significance. There
were strong and statistically significant correlations
between areal exposure of landmarks and their recall, path
exposure and recall, and social significance and landmark
recall. Implications for design are considered.

1340. Vigier, F. Experimental approach to urban design.
        Journal of the American Institute of Planners,
        1965, 30, 21-31.

    An exploratory study of visual search patterns for
tachistoscopically presented photos of urban scenes.
Streets and squares were found to yield quite different
recognition patterns, and differences were observed between
subjects with and without design backgrounds.

1341. Wong, K. Y. Maps in minds: An empirical analysis.
        Environment and Planning A, 1979, 11, 1289-1304.

    The findings of a cognitive mapping study in Hong Kong
were generally in accord with research conducted in Western
cities. Respondents had a strong inclination toward
sequential maps and significant relationships with sex and
socioeconomic variables were observed.

1342. Zannaras, G. An analysis of cognitive and objective
        characteristics of the city: Their influence on
        movements to the city center. Dissertation
        Abstracts International, 1974, 34-B, 3853.

See also: citations 116, 117, 126, 127, 135a, 140, 145, 155, 161, 164, 165, 177 184, 185, 199, 201, 230, 241, 271, 607, 923.

13. INTERIOR ENVIRONMENTS

This final scale-specific chapter focuses on the relationships between human behavior and the size, shape, color, arrangement, equipment, and similar attributes of interior environments. The materials included in this chapter deal with enclosed spaces serving a variety of functions and have been organized into two sections, the first relatively focused and the second quite diverse in content.

a. Office Environments. While, as noted in the Introduction, studies dealing with work environments from a human factors perspective have not been included in this volume, some of the relatively recent research on office environments is seen as more central to environment-behavior studies and is considered here. Much of this work builds upon the concepts of privacy, social contact, interpersonal distancing, and stress developed in Section II. These have included a number of efforts to evaluate the behavioral consequences of "open-plan" and "office landscape" environments, raising many of the same issues as have emerged in studies of open-plan schools (see Chapter 10a).

b. Other Interior Settings. Included here are a number of studies dealing with different attributes of the interior environment. Among the variables considered are room size, shape, color and illumination; the consequences of interior partitioning for flexibility, privacy and social interaction; patterns of pedestrian movement and way-finding behavior within buildings; and the psychological costs of environmental barriers experienced by people with particular disabilites.

a. Office Environments

1343.   Kraemer, Sieverts & Partners, OPEN PLAN OFFICES: NEW
        IDEAS, EXPERIENCE AND IMPROVEMENTS. London:
        McGraw-Hill, Ltd., 1977. 135 pp.

   Based upon evaluations of 20 open-plan offices in West
Germany, findings focus on spatial configuration, patterns
of movement, ambient conditions, and furnishings. The
survey instrument utilized as well as a design check-list
are included.

1344.   Allen, T. J., & Gerstberger, P. G. A field
        experiment to improve communications in a product
        engineering department: The nonterritorial
        office. Human Factors, 1973, 15, 487-498.

   Measurement of employees' attitudes three months before
and eight months after the introduction of an open office
indicated a significant improvement in ease of
communication.

1345.   Bloom, L. J., Weigel, R. G., & Trautt, G. M.
        "Therapeugenic" factors in psychotherapy: Effects
        of office decor and subject-therapist sex pairing
        on the perception of credibility. Journal of
        Consulting and Clinical Psychology, 1977, 45,
        867-873.

   Female therapists occupying traditional offices (i.e.
including professional artifacts and a desk separating
therapist and patient) were perceived as more credible than
those occupying non-traditional settings. Results were
reversed for male therapists.

1346. Boyce, P. R. Users' assessment of a landscaped
office. Journal of Architectural Research, 1974,
3(3), 45-63.

A questionnaire survey was administered to some 300
employees of a utility company prior to, shortly after, and
more than one year after relocation to an open plan office
building. Results focus on both sensory and social
conditions within the new office environment, as well as
primary shortcomings in the areas of heating and
ventilation.

1347. Brookes, M. J. Office landscape: Does it work?
Applied Ergonomics, 1972, 3, 224-236.

Based upon semantic differential ratings, the change
from a conventional bullpen to an office landscape design
resulted in enhanced judgements of aesthetic value, but also
a perceived increase in noise level, visual distraction, and
loss of privacy.

1348. Brookes, M. J., & Kaplan, A. The office environment:
Space planning and effective behavior. Human
Factors, 1972, 14, 373-391.

Reviews current practices in office and space planning
as well as a field study on the effects of office design.
Findings regarding aesthetics, functional efficiency,
privacy and group sociability are considered.

1349. Campbell, D. E. Interior office design and visitor
response. Journal of Applied Psychology, 1979,
64, 648-653.

Students rated 16 photographic slides of a faculty
office illustrating manipulations of spatial arrangement and
decoration. Plants and wall posters led to more positive
ratings of the professor and of being in the office, clutter

led to strong negative ratings, and furniture arrangement
had little impact. Patterns of interaction among variables
and differential impacts upon males and females are noted.

1350. Campbell, D. E. Professors and their offices: A
        survey of person-behavior-environment
        relationships. In #44 (EDRA-11), pp. 227-237.

   Arrangement and decoration of professors' offices were
recorded and related to patterns of office utilization and
significance attached to various design features. In
addition, personality variables (i.e. arousal seeking,
dominance and social competence) were related to office
decoration and seating arrangement.

1351. Duffy F., & Worthington J. Organizational design.
        Journal of Architectural Research, 1977, 6(1),
        4-9.

   A discussion of the implications for architectural
design of such organizational variables as patterns of
growth, size of working groups, relationships between
component parts, distribution of power, and interface with
the public.

1352. Estabrook, M., & Sommer, R. Social rank and
        acquaintanceship in two academic buildings. In
        W. K. Graham & K. H. Roberts (Eds.), COMPARATIVE
        STUDIES IN ORGANIZATIONAL BEHAVIOR. New York:
        Holt, Rinehart & Winston, 1972. Pp. 123-128.

   Studies conducted in two academic office buildings
revealed that, as in other well-established social
structures, communication was directed toward higher status
individuals, even when the propinquity hypothesis would
predict otherwise. The importance of shared, sociopetal
spaces is stressed.

1353. Fucigna, J. T. The ergonomics of offices.
       Ergonomics, 1967, 11, 589-604.

     Activity log data were utilized to compare conventional
and office landscape environments. While efficiency did not
improve in the non-conventional setting, preferences were
expressed for its convenience and superior organization.

1354. Goodrich, R. Office environment post-occupancy
       evaluation: A dialogue. Man-Environment Systems,
       1978, 8, 175-190.

     Reports on an evaluation carried out in the offices of a
major corporation, equipped with innovative, modular work-
stations. Findings are grouped into three sections: worker
efficiency, self-fulfillment and social climate; issues of
perceived privacy, distraction and confidentiality and
impact of the open environment on orientation and
preference.

1355. Ives, R. S., & Ferdinands, R. Working in a
       landscaped office. Personnel Practice Bulletin,
       1974, 30, 126-141.

     The majority of employees relocated from a conventional
to an open plan office perceived communication and
socializing to have improved in the new setting.

1356. Joiner, D. Social ritual and architectural space.
       Architectural Research and Teaching, 1971, 1(3),
       11-22.

     Analysis of layout and use of small offices in various
types of organizations indicated that furnishings and
spatial relationships were utilized to sustain and reinforce
social relationships and to supplement other forms of verbal
and non-verbal communication.

1357.   Justa, F. C., & Golan, M. B. Office design: Is
        privacy still a problem? Journal of Architectural
        Research, 1977, 6(2), 5-12.

Research conducted with 40 middle management executives
suggests that physical supports are necessary but not
sufficient for achievement of privacy. The meaning and
clarity of privacy cues (e.g. partitions) were found to
differ across specific inter-office situations.

1358.   Keighley, E. C. Acceptability criteria for noise in
        large offices. Journal of Sound and Vibration,
        1970, 11, 83-93.

A study conducted in 40 offices indicated that the
perception of a setting as noisy is related to both average
sound level and the number of impact noises rising
substantially above the background sound level. Criteria
for sound control are discussed.

1359.   Lewis, P. T., & O'Sullivan, P. E. Acoustic privacy
        in office design. Journal of Architectural
        Research, 1974, 3(1), 48-51.

A consideration of the conflicting demands of privacy
and communication in the office environment. Attention is
focused on the Articulation Index and the factors which
influence it. A final section deals with designing for
acoustic privacy.

1360.   Lipman, A. et al. Power, a neglected concept in
        office design. Journal of Architectural Research,
        1978, 6(3), 28-37.

Exporatory studies conducted in three office
environments point to the importance of social power
relationships and their implications for organization and
design of physical settings. Given the lack of attention to

such variables in the office environment literature,
implications for research are considered.

@@@ Manning, P. Office design: A study of environment. In
#29 (Proshansky et al., 1970), pp. 463-483.

Reports findings of both preliminary analyses of 21
office buildings in the north-west of England and a detailed
case study of one high-rise in Manchester. Consideration is
given to spatial, visual, thermal and aural features of the
environment.

1361. Nemecek, J., & Grandjean, E. Noise in landscaped
offices. Applied Ergonomics, 1973, 4, 19-22.

Despite the relatively low noise levels recorded in a
sample of 15 office-landscape environments, over one third
of the 519 employees interviewed were severely disturbed by
noise. Content of conversation, rather than its decibel
level, was the most frequently cited disturbance.

1362. Nemecek, J., & Grandjean, E. Results of an ergonomic
investigation of large-space offices. Human
Factors, 1973, 15, 111-124.

An ergonomic investigation of 15 large-space offices
based upon measurements of noise, lighting, and room
climate, as well as interviews with over 500 employees.
Primary advantages over conventional offices included
improved communication and personal contact, while problems
reported were related to concentration and confidentiality.

1363. Oldham, G. R., & Brass, D. J. Employee reactions to
an open-plan office: A naturally occurring quasi-
experiment. Administrative Science Quarterly,
1979, 24, 267-284.

Employees' reactions to work were assessed once before and twice after relocation from a conventional to an open-plan setting. While employee satisfaction and internal motivation decreased significantly after the move, analyses suggest that changes in job characteristics explain much of this decline.

1364. Nichols, K. W. Urban office buildings: View
       variables. In #1078 (Conway), pp. 72-80.

Questionnaire responses provide support for the importance of windows for view as well as daylight. Of those respondents without window views, the vast majority reported making trips to other portions of the building which afforded a view to the outside.

1365. Parsons, H. M. Work environments. In #8 (Altman &
       Wohlwill 1), pp. 163-209.

Includes a review organized in terms of patterns of behavior and salient environmental attributes in work settings.

1366. Preston, P., & Quesada, A. What does your office say
       about you? Supervisory Management, 1974, 19,
       28-34.

Managers were found to arrange their offices with their desks either facing toward the door, reflecting a desire to control communication, or facing away from the door, to remove barriers to communications.

1367. Sloan, S. A. Translating psycho-social criteria into
       design determinants. In #36 (EDRA-3), vol. 1,
       sec. 14.5, pp. 1-10.

A commercial office was redesigned on the basis of both
behavioral observations and characterization of employees in
terms of aggressiveness, sociability, and territorial needs.
Levels of dissatisfaction with the office environment were
reduced by more than 50% after remodelling.

1368. Wells, B. W. P. The psycho-social influence of
        building environment: Sociometric findings in
        large and small office spaces. Building Science,
        1965, 1, 153-165.

Along with age and sex, spatial relationships influenced
sociometric choices of office workers. Fairly close social
groups developed in smaller areas, while social links
connecting people in open areas were less tightly knit.

1369. White, A. G. The patient sits down: A clinical
        note. Psychosomatic Medicine, 1953, 15, 256-257.

Removal of a physician's desk as a barrier separating
him from a sample of cardiac patients led to a five-fold
increase in the number of patients characterized as sitting
"at ease". It is suggested that this enhanced informality
facilitated the physician's interviewing of patients.

1370. Widgery, R., & Stackpole, C. Desk position,
        interviewee anxiety, and interviewer credibility.
        Journal of Counseling Psychology, 1972, 19,
        173-177.

Highly anxious patients perceived a therapist as more
credible when there was no desk separating them, while the
presence of an intervening desk increased the therapist's
credibility for low-anxiety patients.

1371. Zweigenhaft, R. L. Personal space in the faculty
        office: Desk placement and the student-faculty
        interaction. Journal of Applied Psychology, 1976,
        61, 529-532.

Among professors surveyed, 40% placed their desk between
themselves and visitors while 60% used alternative
arrangements which avoided the desk as a barrier. Faculty
who utilized the former configuration were older, of higher
academic rank, and rated less positively on student-faculty
interaction.

See also: citation 506.

b. Other Interior Settings

1372. Bennett, C. SPACES FOR PEOPLE: HUMAN FACTORS IN
          DESIGN. Englewood Cliffs, N. J.: 1977. 195 pp.

     A behaviorally based approach to the design of interior
environments, organized around criteria of health and
safety, task performance, comfort, and aesthetic pleasure.

1373. Lifchez, R., & Winslow, B. DESIGN FOR INDEPENDENT
          LIVING: THE ENVIRONMENT AND PHYSICALLY DISABLED
          PEOPLE. New York: Whitney Library of Design,
          1979. 208 pp.

     Psychosocial interviews with a number of key informants
provide insights into the behavioral consequences of living
in less than totally accessible environments. A Catalogue
presents a variety of user-initiated means of adapting the
home environment to the needs of disabled people.

1374. Acking, C. A., & Kuller, R. The perception of an
    interior as a function of its color. Ergonomics,
    1972, 15, 645-654.

Subjects evaluated slides of room sketches in which the
colors of walls and some room details were varied. Darkness
of the room influenced social desirability, value accounted
for perceived spaciousness, and complexity was found to
depend on chromatic strength of colors; no systematic
relationship, however, emerged between room color and
ratings of pleasantness.

1375. Adelberg, T. Z., & Shelly, M. W. Notes on
    satisfactions in a shopping center, I-IV.
    Psychological Reports, 1967, 21, 507-8
    536; 584; 660.

A series of observational studies of behavior in stores,
with particular emphasis on the arousal value of social and
visual stimuli encountered in such settings.

1376. Baird, J. C., Cassidy, B., & Kurr J. Room preference
    as a function of architectural features and user
    activities. Journal of Applied Psychology, 1978,
    63, 719-727.

Preference ratings were gathered for simulated rooms
varying in ceiling height, ceiling slope, and wall angle, as
well as for variations in presumed user activities.
Preference functions are described as is a quadratic model
for the ceiling height - preference relationship.

1377. Bechtel, R. B. Human movement and architecture. In
    #29 (Proshansky et al., 1970), pp. 642-645.

A presentation of a technique for observing people's
environmental-exploration activity and transversals of
environments by means of the hodometer, which records the

number and locaiton of footsteps across a floor.
Illustrative data from the behavior of museum visitors are
presented.

1378.   Corlett, E. N., Manenica, I., & Bishop, R. P.   The
        design of direction finding systems in buildings.
        Applied Ergonomics, 1972, 3, 66-69.

Way-finding behavior of individuals unfamiliar with a
university building was found to improve markedly when the
existing signage system was replaced with a new one based
upon identification of major destinations and choice points.
Their performance, however, remained much below that of
people familiar with the setting.

1379.   Cunningham, M.   Notes on the psychological basis of
        environmental design: The right-left dimension in
        apartment floor plans.   Environment and Behavior,
        1977, 9, 125-135.

Subjects presented with mirror image pairs of floor
plans demonstrated a strong tendency to select those
apartment plans with the livingroom in the far right, rather
than the far left, corner.   Possible theoretical
interpretations and directions for research are considered.

1380.   Evans, G. W., et al.   Cognitive mapping and
        architecture.   Journal of Applied Psychology,
        1980, 65, 474-478.

The effects of color-coding were tested in the interior
of a large, institutional university building.   The group
which experienced the color-coded condition made fewer
errors in a way-finding task, had higher recall and
recognition memory for building floor plans, and was able to
more accurately locate nonvisible points within the building
with a transit.

1381. Feller, R. A. Effect of varying corridor
       illumination on noise level in a residence hall.
       Journal of College Personnel, 1968, 9, 150-152.

       In three of four test cases, a reduction in illumination
levels in a corridor to 0.5 candlepower effected a
significant reduction in noise levels, in high as well as
low traffic conditions.

1382. Flynn, J. E., et al. Interim study of procedures for
       investigating the effect of light on impression
       and behavior. Journal of the Illuminating
       Engineering Society, 1973, 3, 87-94.

       Several rating scales were utilized to assess six
alternative lighting configurations within a demonstration
room. Results, as well as dimensional analyses of both
ratings and rooms are reviewed. The impact of lighting upon
seating patterns and preferences is also discussed.

1383. High, T., & Sundstrom, E. Room flexibility and space
       use in a dormitory. Environment and Behavior,
       1977, 9, 81-89.

       Flexibility, defined as the degree to which furniture in
a room can be rearranged, influenced residents' ratings of
flexibility, usage of room for inter-personal recreation and
numbers of visitors in women's rooms.

1384. Holahan, C. J. Consultation in environmental
       psychology: A case study of a new counseling
       role. Journal of Counseling Psychology, 1977, 24,
       251-254.

       Design modifications in a student dormitory dining hall
were evaluated by both self-report and observational data.
Provision of partitions in this formerly open space
indicated that privacy was increased, institutional

appearance diminished, and opportunities for social contact
enhanced.

1385. Imamgolu, V. The effect of furniture density on the
       subjective evaluation of spaciousness and
       estimation of size of rooms. In # 50 (Kuller),
       pp. 341-352.

An experimental room was judged as less spacious when it
was empty than when it was appropriately furnished, and
judged as much less spacious when in an over-furnished
condition.

1386. Kasmar, J. V., Griffin, W. V., & Mauritzen, J. H.
       Effect of environmental surroundings on
       outpatients' mood and perception of psychiatrists.
       Journal of Consulting and Clinical Psychology,
       1968, 32, 223-226.

A study of the effects of "beautiful" and "ugly"
consultation rooms on pyschiatric outpatients' self-rated
mood, perceptions of the environment, and ratings of a
psychiatrist. The lack of significant main effects suggests
that the influence of the environment is influenced by
behavior of the psychiatrist and age and sex of patients.

1387. Matson, D. L. Public order in elevators:
       Environmental constraints on public behavior.
       Man-Environment Systems, 1978, 8, 167-174.

Based upon existing research from proxemics and human
engineering, as well as verbally-elicited and observational
data, it is suggested that specific environmental features
within the elevator setting act as constraints upon patterns
of attention.

1388. Melton, A. W. Visitor behavior in museums: Some
       early research in environmental design. Human
       Factors, 1972, 14, 393-403.

Includes slightly edited versions of two papers originally published in 1935 and 1936. Both focus on factors which influence museum visitors' interest in paintings and other exhibits. Factors investigated include position and relative location of art objects, and time intervals for operating displays.

1389. Preiser, W. F. E. An analysis of unobtrusive observations of pedestrian movement and stationary behavior in a shopping mall. In #50 (Kuller), pp. 287-300.

Systematic observation of patterns of pedestrian behavior within an enclosed shopping mall revealed a more leisurely pace of shoppers on soft floor surfaces and under uncrowded conditions. Users of seating areas within the mall preferred to orient themselves toward the major stream of pedestrian traffic.

1390. Rosenbloom, S. Openness-enclosure and seating arrangements as spatial determinants in lounge design. In #40 (EDRA-7), vol. 1, pp. 138-144.

Enclosure of a previously unbounded lounge area created a more private and less populated space, with an increase in study-related as opposed to social activities. Significantly more moving of furniture was observed in the sociofugal than in the sociopetal conditions.

1391. Samuelson, D. J., & Lindauer, M. S. Perception, evaluation, and performance in a neat and messy room by high and low sensation seekers. Environment and Behavior, 1976, 8, 291-306.

Low sensation seekers rated a neat room more positively than did high sensation seekers, while the results were reversed for high scorers. The messy-neat room conditions

III. APPLIED AREAS

led to differential perceptions of the space but had little
impact on the performance of various tasks.

1392.   Spivack, M.   Sensory distortions in tunnels and
            corridors.   Hospital and Community Psychiatry,
            1967 18, 24-30.

It is suggested that many of the visual and spatial
features common to psychiatric facilities are inappropriate
in milieus for emotionally disturbed patients.   The author
reviews a number of his own observations indicating that
excessively long corridors, shiny surfaces and glaring
lights may have negative impacts upon both perception and
behavior.

1393.   Stillitz, I. B.   The role of static pedestrian groups
            in crowded spaces.   Ergonomics, 1969, 12, 821-839.

An exploratory observational study of pedestrian flow in
five London Underground stations.   Static groups (e.g.
people using automatic ticket machines) were found to impede
circulation and diminish the effective width of key movement
channels.

1394.   Tagg, S. K.   The subjective meaning of rooms.   In #47
            (Canter & Lee), pp. 65-70.

Pleasantness ratings were found to be more dependent
upon differences of building type, while spaciousness
ratings were more dependent upon room type.   Patterns of
differences of room meaning were shown to be related to the
range of activities expected to occur within them.

1395.   Tognoli, J.   The effect of windowless rooms and
            unembellished surroundings on attitudes and
            retention.   Environment and Behavior, 1973, 5,
            191-201.

Retention of information presented to subjects on a
video-tape was found to be influenced by the three-way
interaction between presence or absence of a window in the
testing room, the presence of pictures on the walls, and a
hard or soft chair. It is suggested that the impact of
these variables on behavior is the result of a perceptual
synthesis based upon the context in which they occur.

1396. Weisman, G. Way-finding in the built environment:
       An evaluation of architectural legibility.
       Environment and Behavior, 1981, 13 (in press).

Self-report data gathered from a sample of university
students indicated way-finding to be a problem for a
substantial minority of students in particular buildings.
One of four theoretically-derived environmental variables,
judged simplicity of floor plans, was able to account for a
significant proportion of the variance in these way-finding
data.

1397. Wells, B. W. P. Subjective responses to the lighting
       installation in a modern office building and their
       design implications. Building Science, 1965, 1,
       57-67.

While it is widely felt that daylight is important to
people, workers were found to overestimate the contribution
of daylight to illumination levels in their office and to
feel that they had a considerable amount when there was
scarcely any.

1398. Wools, R., & Canter, D. The effect of the meaning of
       buildings on behavior. Applied Ergonomics, 1970,
       1, 144-150.

Reviews the development of a set of semantic scales and
their application in assessing the "friendliness" of line
drawings of rooms varying in window and ceiling design and
seating arrangement. Impacts of rooms on behavior and
evaluations of occupants are also considered.

1399. Miller, M. C. Perceptual foundations of interior
       design. DISSERTATION ABSTRACTS, INTERNATIONAL,
       1971, 32-B, 1005-1006.

See also: citations 108, 112, 119, 120, 149, 220a, 223, 227,
259, 260, 271, 407, 456, 467, 615, 889.

PART IV

COGNATE AREAS

14.   BEHAVIORAL DEMOGRAPHY

Included under this rubric are studies carried out to a large extent by geographers, demographers and others specializing in the study of population. The specific focus of the chapter is that of behavioral and attitudinal research – such as the bases for decisions to move or migrate, or determinants of preferences for different geographic areas as places to live – relevant to issues of population and population distribution. Work based exclusively on demographic data, such as comprises the bulk of research on mobility, migration, population movements, etc., has been excluded from consideration.

1400.   Brody, E. B. (Ed.) BEHAVIOR IN NEW ENVIRONMENTS: ADAPTATION OF MIGRANT POPULATIONS. Beverly Hills, Calif.: Sage Publications, 1970. 479 pp.

A collection of contributions drawn predominantly from sociology, anthropology and psychiatry, dealing with individual and psycho-social adjustment of migrant populations and problems of mental health and deviance associated with such populations.

1401.   Price, D. O., & Sikes, M. M. Rural–urban migration and research in the United States: Annotated bibliography and synthesis. (Center for Population Research Monograph.) Washington, D. C. 1975. (DHEW Publication No. NIH 75-565.) xi, 250 pp.

Contains two chapters of relevance to migration behavior: one on migration decision-making, and one on adjustment of migrants (educational, psychological, social and economic).

1402.   Rossi, P. H. WHY FAMILIES MOVE. Glencoe, Ill.:
         Free Press, 1955. 220 pp.

     A comprehensive investigation of residents of four
sections of Philadelphia, chosen to represent differing
degrees of stability and mobility, to bring out
characteristics differentiating mobile from stable
households and neighborhoods, as well as factors motivating
families to change their place of residence. Data on
residents' awareness of and concern about the mobility or
stability of their neighborhood are also included.

1403.   Brown, L. A., & Moore, E. G. The intra-urban
         migration process: A perspective. Yearbook of
         the General Systems Society, 1970, 15, 109-122.

     Migration behavior is analyzed in terms of two phases,
the first, the decision to migrate, based on a model of
environmental stresses impinging on the individual, and
their resolution through the decision to move, and the
second, the relocation decision, focusing on processes of
environmental evaluation, search behavior, and information
seeking.

1404.   Butler, E. W., Sabagh, G., & Van Arsdel, M. D., Jr.
         Demographic and social-psychological factors in
         residential mobility. Sociology and Social
         Research, 1964, 48, 139-154.

     Studied factors differentiating movers from stayers
(based on expressions of intent to move) among residents of
two Los Angeles neighborhoods. Housing dissatisfaction
showed a stronger association with intent to move than
neighborhood dissatisfaction. The complexity of the role of
socio-psychological determinants of mobility is stressed.

1405.   Butler, E. W., et al. Moving behavior and
         residential choice: A national survey. National
         Cooperative Highway Research Program Report #81.
         Washington, D.C.: Highway Research Board, 1969.
         67 pp.

         A survey of factors relating to residential choice and
preference, including neighborhood satisfaction, type and
style of housing, and role of schools, accessibility and
environmental conditions.

1406.   Clark, W. A. V., & Cadwallader, M. Locational
         stress and residential mobility. Environment and
         Behavior, 1972, 4, 29-41.

         An analysis of individuals' assessment of stresses in
their residential environment and its relation to the desire
to move, and to decisions to move. The latter are treated
as a function of the discrepancy between experienced level
of satisfaction and that perceived to be obtainable by
moving.

1407.   Fuguitt, G. V., & Zuiches, J. J. Residential
         preferences and population distribution.
         Demography, 1975, 12, 491-504.

         Results of a survey of a national-U.S. sample point to a
general preference for residence in smaller cities or towns
or rural locations as a place of residence, but within 30
miles of a large city. Preference is highest for type of
place corresponding to actual residence, but satisfaction
rates increase with decreasing community size, within a
30-mile zone. Differences in the reasons cited for their
preferences are found between those preferring a large city
and those preferring some other location.

1408.   Huff, J. O., & Clark, W. A. V. Cumulative stress
         and cumulative inertia: A behavioral model of the
         decision to move. Environment and Planning A,
         1978, 10, 1101-1119.

A model of decision-making process determining moves on the part of a household is presented, based on the resolution of conflict between cumulative inertia and residential stress. Tests of the model via simulation techniques are shown, and their results compared to actual data on moving.

1409.   Kahoe, R. D. Motivations for urban-rural migration. Journal of Social Psychology, 1975, 96, 303-304.

Studied students' intentions with respect to type of residential location they would choose following graduation. Found a greater incidence of movement away from, as opposed to toward large urban areas. The former movement appeared to be motivated more by a desire to avoid negative aspects of urban communities than to approach positive aspects of non-urban areas.

1410.   Kantor, M. B. Some consequences of residential and social mobility for the adjustment of children. In M. B. Kantor (Ed.), MOBILITY AND MENTAL HEALTH. Springfield, Ill.: C. C. Thomas, 1965. Pp. 86-122.

A review of the literature and report of research on effects of mobility on the psychological adjustment of children, with reference to both changes of residence and of socio-economic status.

1411.   Kegel-Flom, P. Predictors of rural practice location. Journal of Medical Education, 1977, 52, 204-209.

A survey of practicing optometrists residing in rural and urban areas showed two variables, place of origin, and score on the Urbanism scale of McKechnie's Environmental Response Inventory (cf. #98), when used in combination, to effectively differentiate the rural from the urban group. A prediction equation incorporating these two variables, when applied to a group of optometry students, predicted both the students' intended and their subsequent actual choice of location for their practice.

1412.   Lee, E. S. A theory of migration. Demography, 1966, 3, 47-57.

Analyzes migration streams in terms of factors associated with place of origin and destination, intervening obstacles, and personal factors. Emphasizes the role of diversity, both of areas (of origin) and of people, in promoting migration, and points to certain characteristics of different types of migration streams, e.g., positive selection among migrants attracted to certain positive attributes of their destination.

1413.   Mangalam, J. J., & Schwarzweller, H. K. General theory in the study of migration: Current needs and difficulties, International Migration Review, 1968, 3, 3-18.

A review of migration theory and of shortcomings of approaches current at the time of writing. Emphasizes sociological analysis, focused on aggregate phenomena, as opposed to psychological analysis, focused on individual behavior.

1414.   Mazie, S. M., & Rawlings, S   Public attitude towards population distribution issues. In S. M. Mazie (Ed.), POPULATION, DISTRIBUTION, AND POLICY. Commission on population growth and the American Future, Research Reports, Vol. V. Washington, D. C.: U. S. Government Printing Office, 1972. Pp. 599-615.

A survey of data on residential preference, in relation to childhood and current place of residence, and on opinions concerning issues of population distribution.

1415.   Ritchey, P. N. Explanations of migration. Annual
        Review of Sociology, 1976, 2, 363-404.

Although primarily devoted to literature on labor
mobility and social demography, this review includes brief
treatments of decision-making processes in moving, and of
cognitive-behavioral approaches.

1416.   Roseman, C. C. Migration as a spatial and temporal
        process. Annals of the Association of American
        Geographers, 1971, 61, 589-598.

Distinguishes between partial and total displacement
migrations (e.g., intra-urban moves that leave unchanged
much of the daily or weekly movement of the individual
through the environment, vs. moves to a new geographic
region). Correlated differences in information-gathering
processes and the migrants' assimilation to their new
environment are brought out.

1417.   Schultz, R., Artis, J., and Beegle, J. A.
        Measurement of community satisfaction and decision
        to migrate. Rural Sociology, 1963, 28, 279-283.

A study of factors relating to desire to migrate on the
part of high-school seniors in a rural Michigan county
experiencing a high rate of out-migration showed a
significant relationship between such desire and lack of
satisfaction with the community, particularly among girls.
Occupational and educational goals were not related to
desire to migrate.

1418.   Sell, R. R., & DeJong, G. F. Toward a motivational
        theory of migration decision making. Journal of
        Population, 1978, 1, 313-335.

An account of theory and research on migration decision-making, focusing on motivational factors and expectancy of goal-achievement.

1419.   Sonnenfeld, J. Community perceptions and migration
         intentions. Proceedings of the Association of
         American Geographers, 1974, 6, 13-17.

Report of a study of high-school students in seven small Texas towns, comparing those who intended to move away after graduation from high school (movers) with those who did not (stayers). Various indices of community evaluation and landscape preference showed few consistent differences between these two groups (though some important sex differences). Movers did differ from stayers on measures of an ideal-town map.

1420.   Speare, A., Jr. A cost-benefit model of rural-to-
         urban migration in Taiwan. Population Studies,
         1971, 25, 117-131.

Presents and tests a cost-benefit model of the decision to move, including factors of expected change in income, estimated cost, employment status, information concerning job availability, home ownership, and location of relatives. The model was tested on migrants to a Taiwan city moving from surrounding counties, compared to similar residents in those counties. Moderate support for the model's validity was found, though little evidence of conscious weighing of costs and benefits by the mover.

1421.   Speare, A., Jr. Residential satisfaction as an
         intervening variable in residential mobility.
         Demography, 1974, 11, 173-188.

Develops and tests a model of residential mobility placing primary emphasis on residential satisfaction as a controlling variable, and relating satisfaction to such variables as duration of residence, home ownership and crowding.

1422.   Taylor, R. C. Migration and motivation: A study of
         determinants and types. In J. A. Jackson (Ed.),
         MIGRATION. Cambridge (England): Cambridge
         University Press, 1969. Pp. 99-134.

    Differentiates among different motivations for moving
(i.e., those resulting from external forces, those
reflecting aspirations or ambitions of the family, and those
based on separation from friends and relatives or weakening
of personal ties to the home community), and illustrates
these with data from migrants to five villages in England.
The decision-making process, and the problems of adjustment
for migrants are also discussed.

1423.   Van Arsdol, M. D., Sabagh, G., & Butler, E.
         Retrospective and subsequent residential mobility.
         Demography, 1968, 5, 249-267.

    A one-year longitudinal survey of residents of the Los
Angeles area, to determine relationship between actual
moving behavior and prior expressed plans for moving, as
well as previous history of mobility.

1424.   Wolpert, J. Behavioral aspects of the decision to
         migrate. Papers of the Regional Science
         Association, 1965, 15, 159-169.

    A theoretical analysis of the decision-process involved
in migration, based on a combination of "place utility"
considerations (i.e., benefits associated with a particular
locale) and field-theoretical conceptions of the life-space.

1425.   Zuiches, J. J., & Fuguitt, G. V. Residential
         preference: Implications for population
         redistribution in non-metropolitan areas. In S.
         M. Mazie (Ed.), POPULATION, DISTRIBUTION AND
         POLICY. Commission on Population Growth and the

American Future, Research Reports, Vol. V.
Washington, D. C.: U. S. Government Printing
Office, 1972. Pp. 617-630.

Report of a survey of residents of Wisconsin, concerning
residential preference for communities of different types,
stratified both by community size (from metropolitan area to
rural) and by distance from a large city. Results show a
general preference for living in smaller communities or
rural areas, but within a 30-mile distance from a large
city.

1426.   Zuiches, J. J., & Fuguitt, G. V. Public attitudes
         on population distribution policies. Growth and
         Change, 1976, 7(2), 28-33.

A national survey of opinions concerning issues of
population distribution and policies affecting them, with
particular reference to urban growth. Reports data on views
towards controlling vs. encouraging urban growth, and
towards specific relevant governmental actions, as a
function of place of current and preferred residence. Found
overall support for diverse governmental action programs.

1427.   Zuiches, J. J., & Rieger, J. H. Size of place
         preferences and life cycle migration: A cohort
         comparison. Rural Sociology, 1978, 43, 618-633.

Documents the increasing preference for rural areas as a
place of residence on the part of youth from a rural
Michigan region, including both those currently residing in
the area and those who had moved to an urban area.

1428.   Stough, R. R. The cognitive-behavioral approach in
         geography: A theoretical and methodological
         reformulation with a test in the context of intra-
         urban migration. Dissertation Abstracts
         International, 1978, 39-A 2544-2545.

See also: citations 162, 191, 215, 383, 1198, 1202, 1215, 1217.

15. THE ENVIRONMENTAL SOCIAL SCIENCES

While the focus of this bibliography has been on individual behavior in its relationship to the physical environment, the close association between the environment-and-behavior field and a variety of other social-science disciplines should be apparent from an examination of virtually any of the preceding chapters. These contain numerous contributions by geographers and sociologists, as well as a scattering from the fields of anthropology, economics and political science. Some of these fields, notably sociology and economics, have in fact spawned "environmental" specialities akin to that of environmental psychology, but directed predominantly at environmental concerns and environmental-quality issues touching on their respective disciplines. It should thus be helpful to list a few core references from each of these fields, to introduce the researcher or student who might wish to go beyond the psychologically-oriented literature to the major concepts, theories and issues that these cognate fields have developed in their own approaches to problems of man-environment relations.

We have confined ourselves to four or five major references for each cognate discipline, these citations being primarily monographs or readers that present broad overviews of their fields. A few journal articles are also included; they provide either major literature reviews or integrative analyses of the conceptual structure and theoretical issues within their field.

a. General Environmental Social Science

1429.   Carvell, F., & Tadlock, M. (Eds.), IT'S NOT TOO
         LATE. Beverly Hills, Calif.: Glencoe Press,
         1971. xxi, 312 pp.

A reader containing basic articles on environmental
problems and the role of ethical values, historical forces,
political institutions and technology as contributors to
them.

1430.   Duncan, O. D. From social system to ecosystem.
         Sociological Inquiry, 1961, 31, 140–149.

    Considers the interrelationship between social systems
and ecosystems, in terms of mutual dependencies among four
elements: population, organization technology and
environment, of which the first three constitute the "social
complex" proposed by Park. Duncan enlarges that concept
into that of the "ecological complex," incorporating the
physical environment into the system traditionally dealt
with by sociologists.

1431.   Duncan, O. D. Social organization and the
         ecosystem. In R. E. L. Faris (Ed.)., HANDBOOK OF
         MODERN SOCIOLOGY. Chicago: Rand McNally, 1964.
         Pp. 36–82.

    A more extended treatment of the issues dealt with in
the preceding item.

1432.   Dunlap, R. D. (Ed.), Ecology and the social
         sciences: An emerging paradigm. American
         Behavioral Scientist, 1980, 24, 3–151.

    A set of papers dealing with theoretical paradigms and
perspectives from different disciplines on ecological
issues, including anthropology, political science, economics
and sociology.

1433.   Garnsey, M. E., & Hibbs, J. R. (Eds.) SOCIAL
         SCIENCES AND THE ENVIRONMENT. Boulder, Colo.:
         University of Colorado Press, 1968. 249 pp.

A set of essays relating contributions from diverse social sciences -- including economics, political science and geography -- to the study of environmental problems.

1434. Klausner, S. Z. (Ed.), Society and its physical environment. Annals of the American Academy of Political and Social Science, 1970, 389. 187 pp.

A collection of miscellaneous papers on behavioral and social science aspects of environmental issues, including contributions from psychology, sociology, economics, law and political science.

1435. DiCastri, F. International, interdisciplinary research in ecology: Some problems of organization and execution. The case of the Man-and-the-Biosphere (MAB) Programme. Human Ecology, 1976, 4, 235-246.

A discussion of the problems of broad-based interdisciplinary research on environmental problems, as exemplified in UNESCO'S Man and the Biosphere Program. Stresses the need for a problem-focused approach, for the selection of broad units of analysis incorporating human use systems, and collaboration among natural and social scientists in the planning and execution of research.

1436. Klausner, S. Z. Some problems in the logic of current man-environment studies. In #1449 (Burch et al.), pp. 334-364.

An examination of the conceptual problems facing social-science research and theory dealing with problems of the physical environment. Focuses on the non-compatibility and partial incompatibility between the concepts of the environmental scientists, and on the need for a social-scientific language for environmental studies.

See also: citations 4, 87.

b. Anthropology

1437.   Bennett, J. W.   THE ECOLOGICAL TRANSITION: CULTURAL
        ANTHROPOLOGY AND HUMAN ADAPTATION.   New York:
        Pergamon Press, 1976.   378 pp.

    A systematic treatment of human ecology, with particular
attention to differences and relationships between
ecosystems and cultural systems, and the role of adaptation
processes at varying levels of societal development.

1438.   Hawley, A. H.   HUMAN ECOLOGY: A THEORY OF COMMUNITY
        STRUCTURE.   New York: Ronald, 1950.   xvi, 456 pp.

    A classic work on human-ecology theory.

1439.   Helm, J.   The ecological approach in anthropology.
        American Journal of Sociology, 1962, 67, 630-639.

    A review of empirical and theoretical work in cultural
ecology, including antecedents in archaeology and
ethnography.

1440.   Montgomery, E., Bennett, J. W., & Scudder, T.   The
        impact of human activities on the physical and
        social environments: New directions in
        anthropological ecology.   (A symposium.)   Annual
        Review of Anthropology, 1973, 2, 27-61.

A set of three papers covering ecological aspects of health and disease, i.e., their distribution across different cultures, effects of agricultural production on ecosystems, and impacts of major projects of development and technology on human settlements and communities, as well as resulting resettlement of populations.

1441.   Rapoport, A. (Ed.), THE MUTUAL INTERACTION OF PEOPLE AND THEIR BUILT ENVIRONMENT: A CROSS-CULTURAL PERSPECTIVE. The Hague: Mouton, 1976. xv, 505 pp.

A collection of papers on the relationship between culture and environment, with particular reference to housing, settlement patterns, and problems of resettlement.

1442.   Steward, J. H. THEORY OF CULTURE CHANGE: THE METHODOLOGY OF MULTILINEAR EVOLUTION. Urbana, Ill.: University of Illinois Press, 1955. 244 pp.

A systematic treatment of cultural ecology and adaptation to ecological conditions.

1443.   Vayda, A. P. (Ed.), ENVIRONMENT AND CULTURAL BEHAVIOR: ECOLOGICAL STUDIES IN CULTURAL ANTHROPOLOGY. Garden City, N.Y.: Natural History Press, 1969, xvi, 485 pp.

A reader consisting of reprinted papers and chapters from books covering human ecology in application to diverse field settings, and conceptual issues.

c. Behavioral Geography

1444.   Burton, I., Kates, R. W., & Kirkby, A. V. T. Geography. In A. E. Utton & D. H. Henning (Eds.),

INTERDISCIPLINARY ENVIRONMENTAL APPROACHES. Costa
Mesa, Calif.: Educational Media Press, 1974. Pp.
100-126.

Examines the structure of the discipline of geography,
considered as an approach to the study of the
interrelationships between man and environment, and
conceptualizes theoretical approaches in terms of subject-
object, interactional and transactional relationships.

1445.   Cox, K. R., & Golledge, R. G. (Eds.), Behavioral
         problems in geography: A symposium. Evanston,
         Ill.: Northwestern University, 1969. Studies in
         Geography, No. 17. 276 pp.

A collection of papers on topics in behavioral
geography, including locational decisions and preferences,
spatial aspects of voting behavior, migration, and mental
maps.

1446.   Golledge, R. G., Brown, L. A., & Williamson, F.
         Behavioral approaches in geography: An overview.
         The Australian Geographer, 1972, 12 59-79.

A review, detailing the historical origins of behavioral
approaches in geography, and examining major contributions
under five headings: decision making and choice behavior
(e.g., for locational decisions); geographic spread of
information and information diffusion; spatial learning and
search; political goegraphy (role of socio-spatial and
geographic factors in political behavior); and environmental
perception and cognition.

1447.   Jakle, J. A., Brunn, S., & Roseman, C. C. HUMAN
         SPATIAL BEHAVIOR: A SOCIAL GEOGRAPHY. North
         Scituate, Mass.: Duxbury, 1976. 315 pp.

Covers such topics as geographical aspects of human
identity, cognition of the spatial environment, decision
making in migration, siting of social institutions, and
administration of space.

1448.   Manners, I. R., & Mikesell, M. W. (Eds.),
         PERSPECTIVES ON ENVIRONMENT. Washington, D. C.:
         Association of American Geographers, 1974. 395
         pp.

A collection of essays commissioned by the AAG's Panel
on Environmental Education, dealing with conceptual and
substantive aspects of the relationship between the
environment and the discipline of geography. Includes
problems of environmental perception and natural hazards,
but emphasizes environmental impacts of diverse human
activities on particular environmental settings.

See also: citation 32.

d. Environmental Sociology

1449.   Burch, W. R., Jr., Cheek, N. H., Jr., & Taylor L.
         (Eds.), SOCIAL BEHAVIOR, NATURAL RESOURCES, AND
         THE ENVIRONMENT. New York: Harper & Row, 1972.
         374 pp.

An anthology of papers on attitudinal and social-
institutional aspects of natural resource management,
outdoor recreation, and environmental quality.

1450.   Klausner, S. Z. ON MAN IN HIS ENVIRONMENT. San
         Francisco: Jossey-Bass, Inc., 1971. xiv, 224 pp.

An examination of environmental problems from a
sociological and behavioral perspective.

1451.   Little, R. L. (Ed.), Contributions to environmental
         sociology. (Special Issue.) Western Sociological
         Review, 1977, 8, 1-112.

A collection of miscellaneous papers dealing with such
topics as the mediation of environmental disputes, the
environmental movement, boom towns, and migration.

1452.   Buttel, F. H. Environmental sociology: A new
         paradigm. American Sociologist, 1978, 13,
         252-256.

A brief examination of trends in current environmental
sociology, questioning the appropriateness of attributing a
new paradigmatic status to the work and thinking in this
area.

1453.   Dunlap, R. E., & Catton, W. R., Jr. Environmental
         sociology. Annual Review of Sociology, 1979, 5,
         243-273.

A review of the history and institutional structure of
environmental sociology, considering both sociological
approaches to environmental issues, and other areas of
research, including sociology of the built environment,
social response to natural hazards, social impact
assessment, and problems of energy and resource scarcity.

1454.   Dunlap, R. E., & Catton, W. R., Jr. Environmental
         sociology: A framework for analysis. In T.
         O'Riordan & R. C. D'Arge (Eds.), PROGRESS IN
         RESOURCE MANAGEMENT AND ENVIRONMENTAL PLANNING,
         Vol. 1. New York: Wiley, 1979. Pp. 77-85.

Presents an analytic framework for the field of
environmental sociology, examining the concept of human
ecology in its relationship to environmental systems and
issues; discusses the role of environmental change and
causes of environmental degradation, in the context of a
two-fold classification between natural, modified and built
environments and symbolically mediated as opposed to
immediate effects of environments.

e. Environmental Political Science

1455.   Caldwell, L. K. ENVIRONMENTAL STUDIES.
         Bloomington, Ind. Institute of Public
         Administration, Indiana University, 1967. 4 Vols.

A set of four volumes on political and governmental
aspects of environmental issues, covering: "Political
dynamics of environmental control" (Vol. 1);
"Intergovernmental action on environmental policy: The role
of the states" (Vol. 2); "Politics, professionalism and the
environment" (Vol. 3); and "Research and administration in
environmental quality programs (Vol. 4).

1456.   Caldwell, L. K. ENVIRONMENT: A CHALLENGE TO MODERN
         SOCIETY. Garden City, N. Y.: Museum of American
         History, 1970. xvi, 292 pp.

A general treatment of environmental-quality issues from
a political-science perspective, emphasizing ecological
concepts and their implication for matters of policy and
management.

1457.   Caldwell, L. K., Hayes, L. R., & McWhirter, I. M.
         CITIZENS AND THE ENVIRONMENT: CASE STUDIES IN
         POPULAR ACTION. Bloomington, Ind.: Indiana
         University Press, 1976. xxx, 449 pp.

A set of case studies on political action involving
diverse segments of the community, in such areas as land-
use, coastal-zone and inland water protection, wildlife
preservation, air quality, environmental health, and quality
of urban life.

1458.   Milbrath, L. W., & Inscho, F. R.   The environmental
         problem as a political problem. American
         Behavioral Scientist, 1974, 17, 623-650.

A systematic treatment of the role of political science
in application to environmental issues. Presents a matrix
of problems categorized in terms of particular socio-
environmental problems (e.g., environmental constraints on
growth and development; changes in reward structures) and
simultaneously in terms of particular levels of political
action (e.g., administrative; judicial) and of concerns of
political science (e.g., political economy, policy
analysis).

1459.   Milbrath, L W., & Inscho, F. R. (Eds.), THE
         POLITICS OF ENVIRONMENTAL POLICY. Beverly Hills,
         Calif.: Sage Publications, 1975. 136 pp.

A set of papers on environmental political science,
including the role of political structures in air pollution
control, problems of environmental quality in relation to
development, issues in environmental law, and environmental
politics. (Also published as special issue of American
Behavioral Scientist, 1974, 17, 619-770.)

f. Environmental Economics

1460.   Clawson, M., & Knetsch, J. L.   ECONOMICS OF OUTDOOR
         RECREATION. Baltimore: Johns Hopkins University
         Press, 1966. xx, 328 pp.

Considers problems of recreation demand, resource use,
preservation of recreational quality, and issues of costs,
pricing and economic impact relating to outdoor recreation.

1461.   Fisher, A. C., Krutilla, J. V., & Cicchetti, C. J.
         Alternative uses of natural environments: The
         economics of environmental modification. In #682
         (Krutilla), pp. 18-53.

An economic analysis of the relative values associated
with competing uses of a natural resource, e.g., through
development, mining, etc., as opposed to preservation for
natural recreation.

1462.   Kneese, A. V., & Bower, B. T. (Eds.), ENVIRONMENTAL
         QUALITY ANALYSIS: RESEARCH STUDIES IN THE SOCIAL
         SCIENCES. Baltimore: Johns Hopkins University
         Press, 1972. 408 pp.

A collection of papers, primarily by economists, on
relationship between environmental quality and economic
growth, issues of management related to environmental
quality, and legal and political institutions for
environmental quality management.

1463.   Mills, E. S. THE ECONOMICS OF ENVIRONMENTAL
         QUALITY. New York: Norton, 1978. 304 pp.

A general treatment of environmental economics,
including theoretical as well as policy aspects, and
focusing on problems of air and water pollution and solid
wastes.

1464.   Wingo, L., & Evans, A. (Eds.), PUBLIC ECONOMICS AND
         THE QUALITY OF LIFE. Baltimore: Johns Hopkins
         Press, 1977. xv, 327 pp.

An anthology of papers dealing with issues of quality of
life -- and quality of urban life in particular -- from the
perspective of economics and political science.

See also: citation 237.

16. SPECIAL AREAS OF APPLICATION

This final chapter includes a sampling of material in
two rather different areas which have become foci for
applied work in environmental problem solving, making use of
behavioral science information and thinking. The first
section is devoted to the area of environmental design
primarily at the architectural scale. It is made up largely
of contributions by design professionals with interest and
expertise in the behavioral-science dimensions of
environmental design, and focuses on the nature of the
design process and the effective integration of behavioral
science theory and methods into this process. The second
section is devoted to a relatively new area, that of Social
Impact Assessment, which has sprung into being in direct
response to the mandate of the National Environmental
Protection Act to take account of proposed projects that
would alter the impact of the environment on community well-
being and functioning. Social impact assessment thus
represents a realm for the direct application of
environment-and-behavior principles to the domain of
environmental protection, and environmental decision making
and management.

a. Environmental Design

1465. Broadbent, G. DESIGN IN ARCHITECTURE: ARCHITECTURE
        AND THE HUMAN SCIENCES. New York: Wiley, 1978.
        xiv, 504 pp.

A broadly-based review of much recent work in
architecture and related fields. Includes material on the
nature of problem-solving and the design process,
psychological and social needs of people, and statistics and
related methodologies for decision-making.

445

1466. Chermayeff, S., & Alexander, C. COMMUNITY AND
       PRIVACY: TOWARD A NEW ARCHITECTURE OF HUMANISM.
       Garden City N. Y. Doubleday, 1963. 255 pp.

       After developing a critique of the modern urban
environment, with an emphasis on its lack of privacy,
procedures are presented for the systematic design of
alternative forms of housing more responsive to such human
needs.

1467. Deasy, C. M. DESIGN FOR HUMAN AFFAIRS. New York:
       Wiley, 1974. 183 pp.

       Based upon his own experiences in applying behavioral
design research to actual building projects, the author
presents a program for effecting such collaborative efforts.
Difficulties in the process are considered and a number of
examples and illustrations are provided.

1468. Esser, A. H., & Greenbie, B. B. (Eds.), DESIGN FOR
       COMMUNALITY AND PRIVACY. New York: Plenum, 1978.
       344 pp.

       Based upon an EDRA-6 workshop focused on the translation
of research on such concepts as territoriality, privacy,
personal space, and crowding into recommendations applicable
in the environmental design process. Contributions include
papers on crowding, micro-ecology, residential design, and
urban scale environments.

1469. Gutman, R. (Ed.), PEOPLE AND BUILDINGS. New York:
       Basic Books, 1972. 471 pp.

       A reader including contributions on environmental
influences on health and well-being, the social meaning of
architecture, spatial organization and social interaction,
behavioral constraints on building design, and the
application of behavioral science to design.

1470. Heimsath, C. BEHAVIORAL ARCHITECTURE: TOWARD AN
       ACCOUNTABLE DESIGN PROCESS. New York: McGraw-
       Hill, 1977. 203 pp.

   The three parts of this volume review the social costs
of much current environmental design, present an analysis of
buildings and behavior from a systems perspective, and
discuss behavioral research and design processes.

1470a. Hooper, K. Perceptual aspects of architecture. In
       E. C. Carterette & M. P. Friedman (Eds.), HANDBOOK
       OF PERCEPTION, Vol. 9: PERCEPTUAL ECOLOGY. New
       York: Academic Press, 1978. Pp. 155-189.

   An analysis of architecture in terms of major perceptual
dimensions (space, surface, movement of the individual
through space, and function), emphasizing cognitive-
informational, affective-evaluative, and symbolic responses
to the architectural environment. A concluding section
deals with the design process in perceptual terms.

1471. Jones, J. C. DESIGN METHODS: SEEDS OF HUMAN
       FUTURES. New York: Wiley-Interscience, 1971.
       xvi, 407 pp.

   After developing a general model of the design problem-
solving process, a variety of methods, many rooted in the
behavioral sciences, are presented for resolving each phase
of this process.

1472. Lang, J., Burnette, C., Moleski, W., & Vachon, D.
       (Eds.), DESIGNING FOR HUMAN BEHAVIOR:
       ARCHITECTURE AND THE BEHAVIORAL SCIENCES.
       Stroudsburg, Pa.: Dowden, Hutchinson & Ross,
       1974. 353 pp.

Based upon an early conference exploring design-social
science relationships, the contributions to this volume
focus on emerging issues in architectural theory and
practice, fundamental pyschological processes as they impact
design, and problems of methodology; a topical bibliography
is included.

1473. Mehrabian, A. PUBLIC PLACES AND PRIVATE SPACES: THE
       PSYCHOLOGY OF WORK, PLAY AND LIVING ENVIRONMENTS.
       New York: Basic Books, 1976. xiii, 354 pp.

A popularized treatment of the individual's affective
and behavioral response to environments in different
functional settings. Includes chapters on the "intimate"
environment (that of the body, at the surface, as well as
clothed), and on residential, work, therapeutic, play and
communal settings.

1474. Perin, C. WITH MAN IN MIND: AN INTERDISCIPLINARY
       PROSPECTUS FOR ENVIRONMENTAL DESIGN. Cambridge,
       Mass.: MIT Press, 1970. 185 pp.

Presents an approach for regularizing the infusion of
behavioral science theory and methods into the design
process. Argues that the first task of environmental
design is to improve our understanding of the ways in which
physical settings can preserve and enhance people's sense of
competence.

1475. Preiser, W. F. E. (Ed.), FACILITY PROGRAMMING:
       METHODS AND APPLICATIONS. Stroudsburg, Pa.:
       Dowden, Hutchinson Ross, 1978. xii, 340 pp.

Includes descriptions and examples of a number of
efforts to incorporate behaviorally-based design criteria in
the architectural programming process. Specific chapters
deal with office environments, housing, commercial settings,
and correctional facilities.

1476. Sanoff, H. METHODS OF ARCHITECTURAL PROGRAMMING.
       Stroudsburg, Pa.: Dowden, Hutchinson & Ross,
       1977. 184 pp.

     Presents an overview of decision-making and the
architectural programming process, techniques for gathering
and ordering information about human behavior, methods for
translating behavioral information into design terms, and
examples of various programming projects.

1477. Sommer. R. DESIGN AWARENESS. San Francisco:
       Rinehart Press, 1972. 142 pp.

     Argues for the increased participation of users in the
design and a management of the built environment. Part One
focuses on methods of developing environmental awareness as
a necessary first step toward developing informed and active
user groups. Part Two presents methods for evaluation and
incremental improvement of existing environments.

1478. Wade, J. W. ARCHITECTURE, PROBLEMS AND PURPOSES:
       ARCHITECTURAL DESIGN AS A BASIC PROBLEM-SOLVING
       PROCESS. New York: Wiley, 1977. xii, 350 pp.

     Responding to the recent and dramatic increase in the
quantities of social and behavioral information which must
be integrated into the architectural design process, an
effort is made to clarify the nature of this process and to
begin to develop a language for communicating about design.

1479. Merrill, J. L., Jr. Factors influencing the use of
       behavioral research in design. Dissertation
       Abstracts International, 1976, 37-A, 662.

See also: citations 16, 46, 50, 68, 88, 89, 240, 243, 934.

b. Social Impact Assessment

1480. Christensen, K. SOCIAL IMPACTS OF LAND DEVELOPMENT:
       AN INITIAL APPROACH FOR ESTIMATING IMPACTS OF
       NEIGHBORHOOD USAGE AND PERCEPTIONS. Washington,
       D. C.: The Urban Institute, 1976. 144 pp.

   Outlines an approach to obtaining both baseline data on
social conditions in neighborhoods faced with change from
some proposed project, and estimates of changes in such
conditions to be anticipated from the project, relying on
both objective indices (e.g., of crime rates, use of
community resources) and measures of perceptions and
satisfaction.

1481. Finsterbusch, K., & Wolf, C. P. (Eds.), METHODOLOGY
       OF SOCIAL IMPACT ASSESSMENT. Stroudsburg, Pa.:
       Dowden, Hutchinson, and Ross, 1977. 400 pp.

   Includes a section on methodological approaches, along
with sections devoted to more specific aspects of the SIA
process, such as profiling of environmental sites or
conditions, projecting of impacts, assessment, and
evaluation.

1482. Jain, R. K., & Hutchings, B. L. (Eds.),
       ENVIRONMENTAL IMPACT ANALYSIS: EMERGING ISSUES IN
       PLANNING. Urbana, Ill.: University of Illinois
       Press, 1978. xii, 241 pp.

   A collection including a variety of papers dealing with
social impacts, the process of social-impact assessment, and
the utilization of social-impact information in the writing
of environmental impact statements.

1483. Wolf, C. P. (Ed.), SOCIAL IMPACT ASSESSMENT. In #41
       (EDRA-5). Vol. 2. 144 pp.

Includes a review of the state of the art of social
impact assessment by the editor, along with papers on
diverse specific impacts, such as from highway construction,
nuclear power plants, strip mining, etc.; each paper is
followed by a comment and a reply to the comment.

1484. Wolf, C. P. SOCIAL IMPACT ASSESSMENT OF
       TRANSPORTATION PLANNING: A PRELIMINARY
       BIBLIOGRAPHY. Monticello, Ill,: Vance
       Bibliographies, 1979. 35 pp. (Public
       Administration Series, P-250)

Unannotated.

1485. Wolf, C. P. (Ed.), Special issue on social impact
       assessment. Environment and Behavior. 1975, 7,
       259-404.

Includes an overview of the field as well as a series of
case studies on the impacts of water resource management,
coal extraction, new towns, a proposed highway in an inner-
city location, and a public facility in a congested urban
area.

1486. Wolf, C. P. URBAN IMPACT ASSESSMENT: A PRELIMINARY
       BIBLIOGRAPHY. Monticello, Ill.: Vance
       Bibliographies, 1979. 65 pp. (Public
       Administration Series, P-251)

Unannotated

1487. Catalano, R., Simmons, S. J., & Stokols, D Adding
       social science knowledge to environmental decision
       making. Natural Resources Lawyer, 1975, 8, 41-59.

A careful analysis of the California Environmental
Quality Act, and of court decisions relating to its
interpretation and application, to bring out opportunities
for social-science impact into the writing of Environmental
Impact Reports. Special attention is given to the potential
role of environmental psychology in this area.

1488. Cortese, C. F. (Ed.),   The social impacts of energy
        development in the West. Social Science Journal,
        1979, 16(2), 1-127.

A collection of papers on diverse aspects of the social
impacts of energy development, including problems of
employment, out-migration, conditions of residential living
and Navaho-Indian society.

1489. Daneke, G. A., & Priscoll, J. D. Social assessment
        and resource policy: Lessons from water planning.
        Natural Resources Journal, 1979, 19, 359-375.

The water resource planning process is analyzed as a
model for social and life-quality accounting systems
applicable to a broad range of problems of environmental
policy related to resource use.

1490. Jobes, P. C. Problems facing social scientists
        participating in environmental impact research.
        Human Factors, 1977, 19, 47-54.

A general discussion of social scientists' contribution
to environmental impact statements, in terms of
institutional problems, the relationship between social
scientist and client, the relevance of social-science
theory, etc.

1491. Murdock, S. H. The potential role of the ecological
        framework in impact analysis. Rural Sociology,
        1979, 44, 543-565.

Suggests a basic congruence between human ecological theory and the foundation for social impact analysis. The human-ecology perspective is discussed in its implications for both research on and applied aspects of impact assessment.

AUTHOR INDEX

This index includes names of all authors and co-authors of citations in this bibliography, as well as editors and co-editors of monographs, reports, symposia, special issues and the like. Names of authors of papers included in the anthologies itemized in Chapter 1 are also included. Names of institutions or organizations issuing reports are omitted from this index. Numbers refer to citations, not to pages.

Acheson, J.M. 498, 499
Acking, C.A. 242, 1374
Acredolo, L.P. 141
Adams, B. 1106
Adams, R.L.A. 765
Adams, S.K. 904
Adelberg, B. 614, 1255
Adelberg, T.Z. 1375
Agron, G. 14
Ahlbrandt, R.S. 1200
Aiello, J.R. 11, 404,
    405, 406, 529, 530,
    549, 1075
Alban, L. 569
Albert, S. 531
Alexander, C. 1466
Alexander, F. 780
Alfert, E. 1152
al-Hoory, M.T. 751
Allen, G.L. 142, 143
Allen, N.J.R. 639
Allen, P.R.B. 500
Allen, T.J. 1344
Allott, K. 1208
Althoff, P. 783
Altman, I. 8, 9, 10, 11,
    13, 14, 56, 57, 388,
    389, 472, 501, 502,
    506, 516, 881, 1094

Ambrose, K. 939
Ambrose, L. 939
Amick, D.J. 1095
Amir, S. 885
Anderson, H.E. 1002
Anderson, J. 41
Anderson, L. 292
Anderson, L.M. 650
Anderson, T.W. 679
Andrews, H.F. 144
Angell, M.F. 1003
Angrist, S.S. 1096
Ankele, C. 1153
Annis, A.B. 448
Appley, M. 286
Appleyard, D. 9, 89, 145,
    1242, 1315, 1318
Arbuthnot, J. 784, 858, 859
Archea, J. 35
Argyle, M. 532
Arthur, L.M. 634, 640, 641
Artinian, V.A. 940, 941
Artis, J. 1417
Artz, L.M. 874
As, A. 376
Asch, J. 785
Ashcroft, N. 390
Athanasiou, R. 1097
Atwood, B.G. 631

455

464   AUTHOR INDEX

Shepard, M.E. 883
Sherman, M. 384
Sherman, R.C. 207
Sherrod, D.R. 264
Shore, B.M. 785
Shokrkon, H. 1248
Shure, M.B. 971
Shurley, J.T. 380
Siegel, A.W. 10, 142, 143, 173, 174, 208, 209, 210
Siegel, J.M. 363
Sikes, M.M. 1401
Silverstein, M. 1151
Simmons, S.J. 1487
Simon, B.B. 1188
Simpson, C.J. 841
Sims, B. 1339
Sims, J.H. 33, 776
Sims, W.R. 1068
Singer, J.E. 12, 13, 288, 297, 337, 920, 1265
Sinnett, E.R. 592, 1165
Sisler, G. 387
Skanes, G.R. 386
Sladen, B. 418
Slater, B. 972
Sleznick, G. 705
Sloan, S.A. 1367
Slovic, P. 777, 778
Smith, D.E. 497, 1069, 1070
Smith, E.J. 461
Smith, M.A.W. 758
Smith, M.B. 13
Smith, P. 973, 974
Smith, R.H. 1133
Smith, R.J. 562
Smith, T.D. 467
Smith, V.K. 698
Smith, W.M. 385
Sobal, J. 403, 583
Sobel, R.B. 563

Sommer, R. 13, 462, 493, 520, 521, 528, 564, 565, 975, 976, 977, 1034, 1035, 1054, 1071, 1153, 1168, 1169, 1352, 1477
Sonnenfeld, J. 78, 232, 265, 674, 1419
Sorensen, J.H. 11, 779, 782
Sorte, G.J. 242
Southworth, M. 126
Spaeth, G.L. 1003
Spear, B.D. 922
Speare, A.Jr. 1420, 1421
Spector, A.N. 164, 165
Spencer, T.J. 227
Spiro, H.R. 477
Spivack, M. 1392
Spohn, H. 1014
Srivastava, R.K. 1036
Srole, L. 1266
Stackpole, C. 1370
Stankey, G.H. 666, 743, 744, 759
Stave, A.M. 353
Stea, D. 129, 130, 152, 160, 211, 522, 1241
Stebbins, R.A. 978
Steidel, R.E. 1134
Steinberg, H.W. 856
Steinitz, C. 607
Stembridge, D.A. 1037
Stern, P.C. 877, 878
Steven, G. 558
Stevens, A. 212a
Stevenson, A. 1089
Steward, J.H. 1442
Stewart, J. 1215
Stewart, W.F.R. 1090
Stillitz, I.B. 1393
Stillner, V. 380
Stires, L. 979